The Digital Revolution

Synthesis Lectures on Emerging Engineering Technologies

Editor
Kris Iniewski, *Redlen Technologies, Inc.*

The Digital Revolution
Bob Merritt
2016

New Prospects of Integrating Low Substrate Temperatures with Scaling-Sustained Device Architectural Innovation
Nabil Shovon Ashraf, Shawon Alam, and Mohaiminul Alam
2016

Compound Semiconductor Material and Devices
Zhaoiun Liu, Tondgde Huang, Qiang Li, Xing Lu, and Xinbo Zou
2016

Advances in Reflectometric Sensing for Industrial Applications
Andrea Cataldo, Egidio De Benedetto, and Giuseppe Cannazza
2016

Sustaining Moore's Law: Uncertainty Leading to a Certainty of IoT Revolution
Apek Mulay
2015

The Digital Revolution

Bob Merritt

ISBN: 978-3-031-00901-3   paperback
ISBN: 978-3-031-02029-2   ebook

DOI 10.1007/978-3-031-02029-2

A Publication in the Springer series
SYNTHESIS LECTURES ON EMERGING ENGINEERING TECHNOLOGIES

Lecture #5
Series Editor: Kris Iniewski, *Redlen Technologies, Inc.*
Series ISSN

The Digital Revolution

Bob Merritt
Convergent Semiconductors

SYNTHESIS LECTURES ON EMERGING ENGINEERING TECHNOLOGIES #5

ABSTRACT

As technologists, we are constantly exploring and pushing the limits of our own disciplines, and we accept the notion that the efficiencies of new technologies are advancing at a very rapid rate. However, we rarely have time to contemplate the broader impact of these technologies as they impact and amplify adjacent technology disciplines.

This book therefore focuses on the potential impact of those technologies, but it is not intended as a technical manuscript. In this book, we consider our progress and current position on arbitrary popular concepts of future scenarios rather than the typical measurements of cycles per second or milliwatts. We compare our current human cultural situation to other past historic events as we anticipate the future social impact of rapidly accelerating technologies.

We also rely on measurements based on specific events highlighting the breadth of the impact of accelerating semiconductor technologies rather than the specific rate of advance of any particular semiconductor technology.

These measurements certainly lack the mathematic precision and repeatability to which technologists are accustomed, but the material that we are dealing with—the social objectives and future political structures of humanity—does not permit a high degree of mathematic accuracy.

Our conclusion draws from the concept of Singularity. It seems certain that at the rate at which our technologies are advancing, we will exceed the ability of our post–Industrial Revolution structures to absorb these new challenges, and we cannot accurately anticipate what those future social structures will resemble.

KEYWORDS

Makimoto's Wave, Moore's Law, Singularity, artificial intelligence (AI), artificial emotions (AE), robotics, Industrial Revolution, digital revolution, brain-machine interface, uncanny valley, noosphere, braingate, DARPA

Contents

Acknowledgments

I would like to acknowledge the wonderful support and encouragement of Dr. Tsugio Makimoto, as well as the contributions and assistance of my business partner Mrs. Sherry Garber.

Bob Merritt
January 2016

Introduction

Computer scientist and science fiction author Vernor Vinge predicted four ways that a rapid expansion of machine intelligence could occur:[1]

1. The development of computers that are superhumanly intelligent and "awake."

2. Large computer networks (and their associated software devices) that may "wake up" as superhumanly intelligent entities.

3. Computer/human interfaces may become so intimate and tightly bound to a human that those users may reasonably be considered superhumanly intelligent.

4. Biological science may find ways to improve upon the natural human intellect.

The challenge is that the impact of new technologies can be both beneficial as well as destructive, yet the ability to anticipate that impact is very difficult.

For example, the patents granted to a Swedish chemist in the 1800s demonstrate one classic example of this delicate balance. Scientist Alfred Nobel's work in developing explosives was initially intended for use in his family's mining operations in Africa. Unexpectedly, the destructive force of these new high-powered explosives was later used to cause massive destruction in the Crimean War and subsequent conflicts. Recognizing the worldwide impact and destruction resulting from his invention, Nobel laid the foundation for worldwide recognition to be granted to individuals based on their positive contributions to our human condition.

Another consideration of successfully integrating new technology is the rate at which new information is being discovered and integrated into industries, relative to the amount of time until societies can absorb the possibilities and broader implications of that information.

What does the future hold as new technologies are being developed? There are a number of possible scenarios.

An excellent example of a future vision of the integration of technology is the silent movie *Metropolis*, directed by Fritz Lang in 1927. This fascinating story of a humanoid robot programmed by its evil genius inventor to cause havoc is available on the Internet, set to music from various contemporary artists.

Another classic view of the future is George Orwell's book *1984*. This book was required reading in high school in the 1950s and 1960s. First published in 1949, Orwell presented a dysfunctional society in which government employees were continuously re-writing or eliminating

---

[1]http://mindstalk.net/vinge/vinge-sing.html

historical references that were not supportive of the current political goals. This book also introduced the idea of governments waging continuous warfare against vaguely defined and ever-changing external enemies in order to maintain tight control of the society, while distracting the public's attention from the slow decline in their quality of life and the growing encroachment of government into their private lives and personal freedoms.

In the book and 1968 movie *2001: A Space Odyssey*, directed by Stanley Kubrick and co-written with Arthur C. Clarke, HAL, the omnipotent computer, evolved in an unanticipated intellectual direction and eventually wanted to maintain his own level of control. That movie dealt with the elements of an artificial intelligence intent on inserting its own control over that of its human co-worker.

Arnold Schwarzenegger's series of *"Terminator"* movies presents another point of view, with "good" robots battling against "bad" robots for the preservation of mankind.

While each of these scenarios offer different themes for futuristic technologies, this collection of futuristic visions also provides a reference point for evaluating the progress of our current stage of technical evolution. For example, today's level of mobile devices for transferring information has already surpassed the bi-directional yet permanently attached "speakwrite" screen described in the book *1984*. Likewise today's communications channel between computer and human operator exceeds many aspects of the artificial intelligence (AI) interface with HAL in *2001: A Space Odyssey*.

Although it was written from a technologist's point of view rather than as a fictional novel, the book that has arguably comes closest to describing our current condition both technically and socially is *Digital Nomad*, published in 1997 by Dr. Tsugio Makimoto and Mr. David Manners. The concepts and technologies in the book were not described to sensationalize the material but to offer the opinions of a scientist who was deeply involved in the development of advanced technologies.

Dr. Tsugio Makimoto was formerly the chief technology officer at Hitachi Semiconductor and later chief scientist of Sony's robotics program. He is generally recognized as one of the leading engineering talents of Japan and is credited with leading the Organization of Senior Engineers in Japan. He was also recently honored by IEEE as one of the five leading scientists whose contributions have most impacted the electronics industry. Mr. David Manners, a well-known technology commentator and author of numerous articles, remains active as an analyst and commentator of semiconductor events. They published *Digital Nomad* in 1997, which offered a broad view of the new age Internet lifestyle that is now taken for granted. The book was extremely accurate as it envisioned a future business environment supported by high-level professional employees who could reside practically anywhere on the planet. The "work place" became any location from which one's mobile device could achieve access to their personal data.

That Digital Nomad lifestyle anticipated access to three essential elements that were not widely available at that time. These included foremost a relatively inexpensive "Nomadic Toolset," envisioned as portable two-way information devices weighing less than 2 pounds, driven by a high-performance processing element and supported by an adequate amount of operational memory—all combined into a portable device. For all intents and purposes, this "digital toolkit" eventually became today's laptop PCs and tablets.

The next critical ingredient consisted of access to relatively inexpensive high-bandwidth communications at worldwide locations, which of course eventually emerged as today's high-speed Internet.

A final ingredient was unrestricted and affordable travel. The original concept forecasted that professional-level individuals with access to these elements had the option of a nomadic lifestyle if they wished—or a very international network of acquaintances and business partners with whom they could communicate without having to travel at all. The technical conditions for the Digital Nomad lifestyle have thus essentially been met, and to a great extent the lifestyle does exist.

The growth in wireless communications that was a foundation of Digital Nomad was also the critical element in Dr. Makimoto's forecast of a second Digital Wave of semiconductor development that we are now entering. While many in the technology industry were surprised by the decline in demand of PCs and the rise of mobile devices, Dr. Makimoto had predicted this technology crossover several years ago. Dr. Makimoto's other technology forecasts extended past the scenario of the Digital Nomad.

What was unique about Dr. Makimoto's vision was that desktop personal computers were expected to be only the first of several waves of the Digital Revolution. In actuality, the PC wave symbolically crested in the final quarter of 2012. Semiconductor-based Dynamic Random Access Memory, or DRAM, was the primary memory technology used in combination with the processing element for PCs, servers, and large computers. Up to that time, PCs absorbed more DRAM than any other single applications.

However, in the final quarter of 2012, the PC consumption of DRAMs fell to less than 50% of the total DRAM consumption for the first time since the 1980s. That was also the same quarter in which Samsung and other handset manufacturers reported higher than anticipated profitability. Tablets and cell phones as the instruments of choice had become equally necessary items relative to PCs within the Digital Nomad toolkit.

As the primary target for the development of new technologies began to shift, these events foretold a major shift in the technological advances of computing components. The validation of Dr. Makimoto's vision relative to this second digital wave certainly calls for a closer look at the impact and timetable of his later forecast of a third technology wave of the Digital Revolution driven by robotics—particularly when the potential magnitude of that end application is forecasted to exceed the impact of the two previous semiconductor technology waves.

CHAPTER 1

The Next Technology Wave

We are at a historic point in the development of new technologies. In the case of the technologies being presented—just as in many cases in the past—the technical community often knows the most likely path that technologies will follow and what the technologies are generally capable of achieving. Some of the anticipated trends discussed here have been following consistent and predictable paths as forecasted for years.

What technologists don't always know regarding new technologies—and generally haven't been concerned with in the past—is the final commercial form those technologies might take in future applications. Nor have we traditionally been too worried about the political impact, the eventual social compromise, and the cultural integration of the cost/benefits.

However, some long-held theories describing the rate of growth of computing technologies lead to the conclusion that several of these technologies are accelerating fast enough to create substantial social change within the lifetime of most people living today. The concern is that some technical advances are capable of altering our societies faster than can be comfortably assimilated.

The essence of the issue can be captured in two distinct observations that will be presented later in greater detail.

The primary elements of that formula can be stated as:

1. One of the dominant trends in the development of semiconductor technologies associated with computing has been the aggressive rate at which new technologies are being developed. This rate has been recognized as a doubling of those capabilities approximately every two years.

2. As machines, computers have the ability to perform previously defined functions extremely fast. Currently, the ability of a computer to "think" and emulate the human intellect has been estimated to be equivalent to only 0.03% of the human brain.

3. If it is assumed that the intellectual skills of future computing devices continue to progress at the same rate we have seen demonstrated over the past 50 years, mankind should eventually encounter computing devices whose personality is indistinguishable from humans and whose logic and information base exceed that of mankind within the lifetime of the majority of people living today.

Many technologists are in rough agreement with the historic validity of those observations. However, there is strong debate as to whether the trends will continue unchanged into the future—and the degree to which those observations are linked together in the future.

In this book, we are not as interested in the specific measurable rate of advance so much as the rate of advance of those technologies relative to the ability of societies to absorb the implications of those technologies. While the technologies and applications themselves are fascinating, they serve only as the "cause" of our story while our true interest is in their "effect."

Rather than continuing the technical debate over how to precisely forecast the rate of growth of certain technologies, several advancing technological trends in the general area of computers and robotics are presented at a very high level.

We select two active development paths that exemplify technical objectives beyond which we cannot anticipate the absorption of those concepts into every-day life. We have identified the current status of those activities and have considered the possibility of the social impact of what those technologies may ultimately achieve within the lifetime of most readers.

Both of these applications offer solid measurable points of their current status as well as easily identifiable criteria for measuring future developments. In both cases, progress in the development along the technology path of this particular technology would require a major shift in our visions of society.

The first example is the merging of humans and machines to the point that a separate distinction is no longer meaningful. Such applications typically include advanced exoskeletons and computing devices driven by human thoughts via direct hardware connections to the human brain.

The second application promises faster intellectual collaboration and development of other new and emerging technologies—while it also suggests the prospect of an accelerated biological evolution of human intellect through direct connections among a number of human brains.

In the second application, we already know that brain-machine interfaces are becoming more efficient. In 2012, a team at Brown University demonstrated a computer that could interpret the commands of a 58-year-old woman who had been paralyzed by a stroke for almost 15 years. The computer is attached to inserts in her brain and controls a mechanical arm. Based on her mental commands, the computer and associated hardware is capable of selecting a bottle of water or preparing a cup of coffee and lifting those items to serve a drink to her or to a visitor. The actions of the mechanical arm are completely under the command of her thoughts and corresponding brain waves. Any Internet search of "BMI" (brain-machine interface) will provide an update of the growing amount of research in this area.

And what about Cathy Hutchinson, the brave subject of that program? Her latest achievements will be presented in a later chapter.

The reality of the physical linkage between machines and human brain activity has therefore already been proven. The next step is to extend the range of commands and complexity of activities. Expanding those activities eventually requires complex mapping of the interior of the human brain in order to find additional specific contact points, and scientists in several countries have programs under way to complete this task.

On a less global scale however, remarkable progress has already been made in connecting prosthesis devices to undamaged nerve endings and bypassing areas that have been damaged. Progress in this area has led to replacement of hands with a rudimentary ability to duplicate the sense of touch and feeling of surface textures. The market area of medical devices will be the proving ground for many of these technologies.

The second technology path leads toward physically linking together multiple human brains to achieve faster collaboration in the development of new and emerging technologies—while also exploring the prospect of an accelerated *biological* evolution of human intellect.

Significant progress is also being made in this area. While the current state of development in humans is at a very basic level to transferring signals over the Internet from one subject to another that control muscle responses, lab experiments have extended to rats the ability to sense ultra-violet lights—as well as possibly establishing a wired transmission channel between two subject rats.

Integrating the wave of future technologies is emerging as a daunting social challenge. However, mankind has been in this situation previously. From a historic point of view, there is one relatively recent example of technologies advancing faster than could be absorbed into social structures. That example is the 50-year core period of the Industrial Revolution.

At the beginning of that industrialization cycle no one living in the feudal social structure could have ever anticipated the technological, cultural, and political changes that lay ahead, and we will review some of the challenges associated with that period of rapidly advancing technologies.

Based on the accelerated rate of development of new technologies, we are rapidly approaching another similar transitional period. We still don't know the commercial forms that the new technologies might take or how difficult it will be to integrate them into our societies.

However, the difference this time is that we are forewarned of the challenges ahead.

CHAPTER 2

Makimoto's Technology Waves

In December 2013, *Computer Magazine*, which is published by the IEEE Computer Society, devoted a section to the review of "some of the most popular computing laws introduced in the past century and to see where they stand now and what they tell us about the future."

Five specific observations were chosen for the "transformative impact on the essence and appeal of computing, each—with its particular technical focus—helps our rapid progression from giant, multi-ton government machines to the Internet of Things (IoT), digital nomads, and wearable computers."[1]

Two of these computing laws are tangential to the observations in this book, while the other three laws are directly relevant.

The three computing laws that have the most impact on our topic are Metcalfe's Law, Moore's Law, and Makimoto's Wave.

Metcalfe's Law, from Bob Metcalfe, states that the value of a network grows as the square of the number of its users. While difficult to quantify, the law explains the motivation and attraction of increasing the number of users on any particular network. This concept is also the driving force behind increasing the user base of subscribers for applications such as e-mail and other online services.

Moore's Law, first published in 1964, predicts that the chip capacity for logic circuitry doubles at a predictable rate. That rate has slowed somewhat over the past 50 years, but the observation has been the driving power behind the growth in semiconductors and computers. In practice, it meant that if you consider the laptop computer in your hand, the version available at a similar size and price in approximately two years would likely contain twice as much circuitry and processing capability.

More recent surveys indicate that many semiconductor business leaders believe that the performance advances implied by Moore's Law might slow somewhat in the future. However, there is no thought that some insurmountable technical challenge is waiting ahead.[2]

Makimoto's observation was more relevant to the original equipment manufacturers (OEMs). Makimoto's Wave pointed out the predictable 10-year swing of technology development and manufacturability between standardization and commoditization. The standardization cycle encourages manufacturing efficiencies, cost containment, and growth in market share while the opposite cycle of commoditization leads toward competition based on product differentiation, increased performance, and decreased power consumption.

---

[1] http://www.ask2know.net/blog/2014/02/16/computing-laws-origins-standing-and-impact/
[2] http://www.eetimes.com/author.asp?section_id=36&doc_id=1325641

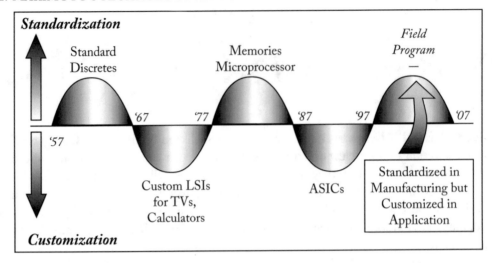

Figure 2.1: Makimoto's Wave. Source: Dr. Tsugio Makimoto, Sony Corporation, paper titled "The Hot Decade of Field Programmable Technologies," published in 2002.

Though many technologists have presented their opinion of the future impact of advanced technologies, we want to specifically consider other forecasts from Dr. Makimoto.

His point of view also separated the Digital Revolution itself into separate waves of technology development with a corresponding impact on societies and businesses.

The IEEE International Electron Devices Meeting (IEDM) is one of the largest international conferences on semiconductor process and device technology. Every year at IEDM, three plenary talks, one each for the European, U.S., and Asian areas, are given as keynote speeches. The subjects of these talks are chosen from the technological trends around the world. On the first day of the IEEE-sponsored conference of 2002, Sony's Dr. Tsugio Makimoto presented the Asian area plenary talk, titled "Chip Technologies for Entertainment Robots—Present and Future." Dr. Makimoto had prepared his presentation in conjunction with Sony's Dr. Toshitada Doi (corporate executive vice president), the originator of Sony's robot development program. The over riding message of his forecast was that at some point soon after 2010, the market influence of digital PCs would be exceeded by that of a different technologically driven product, and the market significance of PCs would decline. Propelled by digital consumer products and data network infrastructure, the forecasted second wave was anticipated to overtake PCs in 2012–2013 and that new wave would support an even larger contribution to the Digital Revolution.

We have already experienced the social and economic impact of PCs, and we are now seeing the shift of market strategies as PCs have yielded to the second technology wave of digital consumer devices and mobile networks.

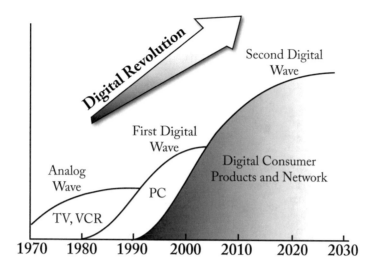

Figure 2.2: The Digital Revolution. Source: Dr. Tsugio Makimoto, Sony Corporation, paper titled "The Hot Decade of Field Programmable Technologies," published in 2002.

A further application of his wave theory relative to end markets identified the primary commercial application of new technologies. His previous outlook had forecasted the transition from PCs to digital consumer and network applications. However, his more recent outlook forecasted the rising wave of robotics as the future target application for the development of new semiconductor technologies.

This particular chart is startling for two reasons. The first reason is the prescience at which he anticipated the timing of the decline in PCs as the driving force for new technologies and the emergence of a second digital wave driven by consumer and networking applications.

The second reason is the anticipation of an equally disruptive technology transition that is still to come.

The Digital Revolution does not end as the market influence of PCs declines or as the cycle of Moore's Law slows from 2 years to 2.5 years. We are only now just beginning to ride this new second wave of networks and mobility—while another wave driven by technologies focused on robotics is still to come.

One of my favorite authors, historian Niall Ferguson, visited the technology development centers of Silicon Valley in California in 2012. He was not impressed by the technical enthusiasm or the futuristic arguments, and he published his reaction.

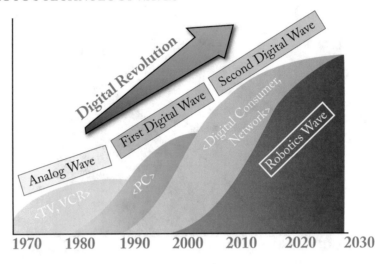

Figure 2.3: The Digital Revolution.

"Are you a technoptimist or a depressimist? This is the question I have been pondering after a weekend hanging with some of the superstars of Silicon Valley. I had never previously appreciated the immense gap that now exists between technological optimism on the one hand and economic pessimism on the other. Silicon Valley sees a bright and beautiful future ahead. Wall Street and Washington see only storm clouds. The geeks think we're on the verge of The Singularity. The wonks retort that we're in the middle of a Depression… My pessimism is supported by a simple historical observation. The achievements of the last 25 years were actually not that big a deal compared with what we did in the preceding 25 years, 1961–1986 (e.g., landing men on the moon). And the 25 years before that, 1935–1960, were even more impressive (e.g., splitting the atom). In the words of Peter Thiel, perhaps the lone skeptic within a hundred miles of Palo Alto: In our youth we were promised flying cars. What did we get? 140 characters.

Moreover, technoptimists have to explain why the rapid scientific technological progress in those earlier periods coincided with massive conflict between armed ideologies. (Which was the most scientifically advanced society in 1932? Germany.)

So let me offer some simple lessons of history: More and faster information is not good in itself. Knowledge is not always the cure. And network effects are not always positive.

I wish I were a technoptimist. It must be heart-warming to believe that Facebook is ushering in a happy-clappy world where everybody "friends" everybody else and we all surf the net in peace (insert smiley face). But I'm afraid history makes me a depressimist. And no, there's not an app—or a gene—that can cure that."[3]

The theme of this book is not the debate about the rate at which new disruptive technologies are being developed, or the division between "technoptimist" and "depressimist," or about the market integration challenges of Makimoto's third wave of the Digital Revolution. The focus of this book is on the social integration of those technologies. Our interest is on the challenge ahead as societies struggle to integrate these future waves of disruptive technologies.

The future impact of new technologies that are continuing to present new challenges as they are integrated into the markets does not depend on technologies continuing to double computing performance infinitely into the future. The theme of this book is that technologies already on our intellectual horizon will be quite sufficient to trigger a substantial change in our social values and political norms.

---

[3]http://www.newsweek.com/niall-ferguson-dont-believe-techno-utopian-hype-65611

CHAPTER 3

The Digital Revolution

Our title of "Digital Revolution" draws attention to the change from analog mechanical and electronic technologies such as color film, table radios, and the heavy telephones of the 50s to the digital forms of technologies that perform those same tasks today.

The revolutionary aspect of this technical change is two-fold. The first impact is the greater cost-efficiency at which information can be obtained and stored by using digital technologies. The storage and processing of analog information is often much more difficult than preserving an approximate measure of that information as digital information.

The second impact is that digital electronic devices can also work at an extraordinarily rapid rate. This particular capability—in conjunction with the reduction in size and power requirement—has resulted in an accelerated pace at which knowledge about our physical world is increasing and is more rapidly being shared with other groups.

This increase in knowledge is derived from the mass production of digital logic circuits, the widespread usage of these devices in computers and mobile communications devices, and the rapid rate at which other groups have access to that new information.

This second element of the Digital Revolution results from the manufacturing efficiencies of digital logic in the form of semiconductor integrated circuits (ICs). In contrast to an analog device with an infinitely broad spectrum of values or positions such as an old AM radio, many common machines work in a digital mode.

In the areas of processing, storing, and transmitting information, this ability to increase efficiencies by utilizing digital technologies has in turn led to a dramatic increase in digital content as well as an accelerated effort to convert more activities into digital functions in order to benefit from these efficiencies—such as digitized music.

In a bigger sense, the term "Digital Revolution" also implies sweeping social and cultural changes brought about by the vast amount of information that is shared over the worldwide Internet.

The social impact of the Digital Revolution results from dealing with the accelerating rate at which knowledge and new information is being generated. As the manufacturing efficiencies of digital technologies increase, the efficiencies of these digital techniques will also increase both the amount of knowledge as well as the broad distribution of that knowledge on a worldwide basis.

This digital content comes from everywhere: sensors used to gather and forecast climate information, posts to social media sites, digital pictures and videos, digital storage of purchase transaction records, and cell phone GPS signals, to name a few.

Another challenge to large corporations is dealing with the huge amount of digital data that is being created. In terms of the amount of data being generated, various technology companies have attempted to forecast the amount of data generated over time. For example, EMC and other companies have estimated that the world created approximately 1.8 zettabytes of data in 2011, and that by 2020 the world will generate 50 times that amount of information.

In addition to the tremendous amount of information being generated, an equally challenging aspect of the Digital Revolution is the ability to transport and store that much information. As the size of the digital universe continues to expand, some sources anticipate that the amount of digital information being stored is more than doubling every two years. The amount of information storage capacity required has been estimated to be almost 8 zettabytes by 2015—which is the equivalent number of bytes as 1,600 times the number of all human words ever spoken.[1]

The worldwide growth in the ability to generate, transport, and store that much data leads directly to the third challenge—which remains the most critical impact of all. The worldwide structure supporting digital forms of information makes it much more cost-effective to distribute that information to the broader general population. The growing issue is that the rate at which new information is becoming available creates conflicts between those who have embraced certain new concepts versus those who fear that this new information and knowledge will cause cultural conflicts. The issue isn't necessarily related to the technology itself, but results from the concern that the rate of cultural change and social impact is too fast to be comfortably absorbed into some communities.

As new information is being discovered and disseminated over the Internet and other electronic means, that information is becoming widely available at a faster rate than can be integrated into many societies without causing conflicts with existing cultural values.

The sight of people so preoccupied with digital devices that they seem to have lost all sense of their immediate surroundings is also an all too common problem. In response to this danger, lawmakers in most industrial countries have authorized fines for cell phone usage while driving a vehicle. An attachment to a recent bill in California also allows for a lesser fine for riding a bicycle in traffic while using a cell phone. At the same time, many 6-year-old children are also at a high enough level of familiarization with digital devices to be perfectly adept at picking up any laptop computer or cell phone and maneuvering their way through the directory to open their favorite games.

Continuing advances in the manufacturing efficiencies of digital technologies are leading to an exponential growth in knowledge, as well as increasing the complexities of those social disruptions.

---

[1]IDC's Digital Universe Study, sponsored by EMC, June 2011.

The point isn't that technologies are inherently harmful or that social disruptions are inevitable or indicative of the attitudes an entire society. Most societies conclude that the broader benefits of these technology advances eventually outweigh the cultural disruptions in most cases. The cultural challenge is that inevitably some elements of any society can be expected to absorb or take advantage of these technologies at a much faster rate than other groups within that society.

A second characteristic of the Digital Revolution is that the manufacturing efficiencies of technology continue to drive the cost reductions and performance increases to such high levels of efficiency that the advances in technologies and the creation of this much new information are also coming at a constantly accelerating rate. The growth in the number of individuals and groups that cannot comfortably absorb this much new information into their cultural structures is the cultural and social challenge ahead.

While the future rate of progress and knowledge growth is a topic of debate among technologists, we can likely assume that the challenges of integrating future technology advances will continue, and that this rate of growth will also lead to further gaps in the rate at which some cultures and societies can integrate the utilization of new technologies.

Many writers consider the current growth in semiconductor and computing efficiencies as powerful enough to represent the third phase of the Industrial Revolution. In the April 21, 2012, *The Economist* published that:

> "The first industrial revolution began in Britain in the late 18th century, with the mechanization of the textile industry. Tasks previously done laboriously by hand in hundreds of weavers' cottages were brought together in a single cotton mill, and the factory was born. The second industrial revolution came in the early 20th century, when Henry Ford mastered the moving assembly line and ushered in the age of mass production. The first two industrial revolutions made people richer and more urban. Now a third revolution is under way. Manufacturing is going digital. As this week's special report argues, this could change not just business, but much else besides.
>
> A number of remarkable technologies are converging: clever software, novel materials, more dexterous robots, new processes (notably three-dimensional printing) and a whole range of web-based services. The factory of the past was based on cranking out zillions of identical products: Ford famously said that car-buyers could have any color they liked, as long as it was black. But the cost of producing much smaller batches of a wider variety, with each product tailored precisely to each customer's whims, is falling. The factory of the future will focus on mass customization—and may look more like those weavers' cottages than Ford's assembly line."[2]

---

[2]http://www.economist.com/node/21553017

CHAPTER 4

Emergence of the Second Digital Wave

What was the mechanism enabling these technological advances to progress and support the concept of a Digital Nomad?

The capabilities of many digital electronic devices—processing speed, memory capacity, sensors, and even the number and size of pixels in digital cameras—are strongly linked to Moore's Law. All of these technologies can present measurements indicating performance growth at roughly exponential rates. This exponential improvement has dramatically enhanced the impact of digital electronics in nearly every segment of the world economy.

It's therefore not surprising to find in California's Silicon Valley an academic organization, Singularity University, whose "mission is to educate, inspire and empower leaders to apply exponential technologies to address humanity's grand challenges."[1]

Another element is that while the reduction in the size and manufacturing costs of transistors would continue, those advances in manufacturing techniques would eventually extend into other areas not traditionally associated with the 1's and 0's of digital logic. In particular, other formats for information outside of the world of discrete numbers (such as radio signals or colors) that typically relied on analog scales of information could also benefit from these advances in digital technology.

Wikipedia defines an analog signal as one that "contains information using non-quantized variations in frequency and amplitude."

In contrast, a digital signal is defined as "a physical signal that is a representation of a sequence of discrete values."

The essence of Dr. Makimoto's insight was the ultimate market limits to the PC-related growth cycle as well as the eventual rise of mobile communications as the next growth cycles that could also benefit from a digitized format. Advances in manufacturing processes and design techniques would eventually enable a second wave of growth based on consumer products and communication networks within his anticipated timeline.

The rapid growth of the second wave results from the ability to extend the digital electronic techniques to more cost-effective support what had traditionally been analog information.

However, an additional technical challenge arose from the shift to mobile devices. This is related to the nature of the memory used to store digital information. As this new wave

---

[1] http://singularityu.org/

of mobile communications applications appeared on the horizon requiring a different set of cost/performance tradeoffs, new directions in technology development inevitably also appeared.

The driving force for the first digital wave was the predictable rate at which digital processing could be increased. These manufacturing capabilities advanced the computational power and software structure of larger computers down to the smaller scale of desktop and personal computers. When combined with corresponding progress in software development, spreadsheets for data processing and word processors for text editing became common tools that changed the nature of the business workforce.

The second challenge in the progress of these disciplines is that the Digital Nomad lifestyle also depends heavily on the anticipated advances in personal computing capabilities and the expected proliferation of mobile communications with easy access to a worldwide communications network.

What is unique in the present timeframe is that mobile connections, which now outnumber fixed residential telecommunications lines by almost 2 to 1, have also become increasingly dominated by digital communications. The digital communications revolution wave continues to radically change the social landscape as well. Consumer usage of these network services has continued to expand as new services become available. Over 85% of U.S. health consumers now maintain that emails, text messages, and voicemails are as helpful as in-person or phone conversations with health providers.[2] Some studies suggest that the impersonal nature of online questionnaires may actually yield more accurate information than face-to-face discussions on emotionally sensitive medical issues.

This combination of exponential growth in digital processing, digital communications, and the growth of the Internet has broad implications that impact media, literacy, and social structures, while contributing to new forms of aggressive behavior in the form of harassment in various public media platforms such as Facebook or Twitter.

The emergence of mobile communication also had a significant impact on the digital communications device market. Internet-connected devices including smartphones and tablets offer a wide range of applications, as seen in the CAGR of smartphones and tablets over other forms of communication.

As a result, an increasing shift toward non-voice communications has been noted. Ofcom's 2012 Communications Market Report in Europe reported that the average consumer was sending around 50 text messages per week, a figure that had doubled over the previous four years. So much of the information presented is now in visual formats, that some academics have pondered whether today's students need additional visual literary skills to understand information that integrates images, videos, sequences, design, form, symbols, color, 3D, and graphic representations. They

---

[2]http://www.televox.com/newsroom/televox-study-reveals-personalized-virtual-messages-are-key-to-patient-engagement/

Table 4.1: Worldwide mobile device shipments

Category	2014	2018	CAGR 2014-2018
	Unit Shipments Millions	Unit Shipments Millions	
Smart phones	1,282	2,724	22.5%
Basic Mobile Phones	597	45	−46.6%
Notebooks/Ultra/Netbooks	210	169	−3.7%
Tablets	230	266	3.9%
Smart Watch	7	193	104.2%
Other Mobile Products	887	1,079	5.0%

need to know how to interpret visual messages and look beyond the surface to determine deeper meaning in what they see."[3]

In regard to the broader industry impact, tablet PCs and advanced cell phones from suppliers such as Apple and Samsung are expected to exceed the mobile PC market over the next few years and become the primary growth driver for semiconductor content. Tablet shipments are forecasted to surpass notebook shipments in 2016. According to Convergent Semiconductors' research, overall mobile device shipments are expected to grow to over four billion in 2020.

The world of high-tech startup companies is one of constant development and rapid innovation as new categories and startup companies create new market opportunities—such as a video technology that changes how people create and view video for mobile devices. One new application is video content specifically intended for display on mobile devices, and it takes full advantage of the fact that users can move and manipulate the images on their mobile phones. The usage cases range from 360-degree tours of car interiors to the creation of remote-participation experiences for concerts and sports games. The software is designed to give viewers the feeling that they're physically right where the video was being recorded. As viewers turn their phones to the left or right, the perspective and frame of reference shifts with them as if they were behind the camera and directing the view to be recorded.

While the growth of tablets and the proliferation of cell phone applications continue to take market share from PCs, other applications convert laptops into mobile digital "toolkits." GPS and geodetic positioning applications were also early supporters of the ability to enable "augmented reality" applications. Augmented reality (AR) concepts search for specific screening parameters and alter or highlight visual content that meets those parameters. This capability relies on the

---

[3] http://www.press.jhu.edu/journals/portal_libraries_and_the_academy/portal_pre_print/articles/13.1hattwig.pdf

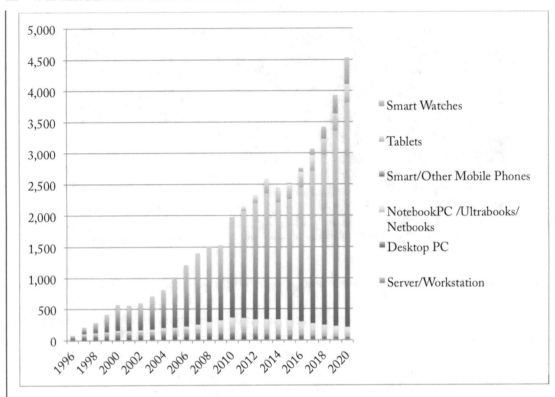

Figure 4.1: Worldwide mobile device shipment forecast (millions). Source: Convergent Semiconductors.

usage of software capable of actively searching the Internet for the particular information that was requested.

Companies such as Hover are already deeply embedded in generating and augmenting 3-D maps of urban areas, and the AR firm Mateio has successfully developed snapping algorithms that pull this kind of data from the cloud and tightly align it to the real world. For example, the image in Figure 4.2 from Mateio represents a mobile phone application that was developed for companies searching for available office space. When linked to the server of a real estate company, the skyline view from a mobile phone or tablet highlights buildings with available office space that fits the search criteria.

Smart glasses have also become available, with options for cameras, microphones, speakers, and other augmented reality displays. These devices are frequently unobtrusive enough to remain outside your pocket and can be positioned to record events as they are developing around you.

While translation capability is already available on laptops and other portable computing devices, future applications are expected to use augmented reality displays to translate and display

Figure 4.2: Source: Mateio.

spoken words as written text. NTT Docomo already demonstrated in 2013 at CEATEC, the annual Japanese electronics trade show, an application using a pair of Vusix smart glasses that sends visual information recorded by the glasses' camera to a server or storage file in the cloud. The glasses can also quickly translate text from one language to another, and quickly send the translation back to the glasses' viewfinder and overlay it onto the document.[4]

Docomo plans to refine the application and target visitors to the 2020 Olympics in Tokyo. When wearing a pair of smart glasses equipped with a translation app, tourists might more easily read menus and signs. As the new wave of digital technologies shifts to mobile digital consumer products, much of the necessary networking technologies are already in place.

---

[4]http://singularityhub.com/2013/10/28/docomo-shows-glasses-that-translate-foreign-languages-right-in-front-of-your-eyes/

CHAPTER 5

Technical Impact

Along with Dr. Makimoto whom we have previously mentioned, Gordon Moore and Robert Metcalfe were also honored by IEEE, recognizing the major contributions to the growth in the computer industry.

Dr. Moore's contribution was associated with his observation of the predictable increase in the performance of microprocessors. In any physical model, the amount of time necessary for any element, such as a car, to go from one point to another can be calculated by knowing the distance to be traveled and the speed at which the physical element is traveling. In the world of digital circuitry, the extremely high speed at which that electrical energy passes through the circuitry remains a constant value at near the speed of light, and the primary variable for performance in completing a task can be computed based on the distance from the beginning of the circuitry path to the end of the physical path. The total elapsed time necessary to complete a computation or retrieve new data is directly related to the physical distance of the route. Therefore, if all of the elements in that route can be made physically smaller, then the amount of time necessary to complete that path can be reduced in the same ratio as the relative comparison of the total distance between two similar circuits. In other words, if the entire construction of the circuitry can be reduced by 50%, then the amount of time required to complete that task can also be reduced by approximately 50%.

One of the significant elements in creating these information pathways is the minimum width at which the metal "lines" can be constructed within that manufacturing process. Intel co-founder Dr. Moore had observed that the progress in making smaller and smaller electrical circuits appeared to occur at a relatively constant rate, as advances in manufacturing techniques appeared to support the possibility of reducing the total area of necessary circuitry by approximately 50% every two years.

In practice, this meant that if you were a PC manufacturer and you looked at the size of the desktop computer on your desk 10 years ago—or the size of the cell phone in your hand today—the manufacturers of those devices could anticipate several options in planning for future products. The manufacturer could anticipate that a re-design of that same product in two years would require only half as much physical space—and therefore a significant reduction on manufacturing costs. Another option that the manufacturer might anticipate is that a device with twice the performance of the current product would also be available two years into the future at roughly the same physical size and materials.

From a marketing perspective, a key concern was whether the end market was growing fast enough to absorb that much of an increase in computing efficiency. If market demand could not absorb the increases in manufacturing efficiencies, then each step forward in manufacturing efficiencies would eventually begin to reduce the size of the total end market.

What social element could create enough market demand to sustain a doubling of performance every two years—or would the market for desktop and personal computing devices simply become saturated?

Fortunately for the industry, there was another factor at play relative to the market size. This shift in market requirements has often been attributed to another significant observation—Metcalfe's Law. Metcalfe's Law was first articulated by George Gilder in 1993 relative to telecommunications networks and attributed to Robert Metcalfe in regard to the potential growth rate of the Ethernet. Metcalfe's observation is expressed in two general ways:

1. The number of possible cross-connections in a network grows as the square of the number of computers in the network increases.

2. The community value of a network grows as the square of the number of its users increase.

Metcalfe's Law is often cited as an explanation for the rapid growth of the Internet. The concept of the increased value of multiple connections also re-appears in a later chapter related to the sharing of information. While Moore's Law predicts the rate at which computer power is accelerating, Metcalfe's Law explains the motivation for the rising wave of information technology and the increased value of broad-based networks.

We've already experienced how the Internet changes the way we work and maintain social contacts. However, the value of a network increases exponentially as the square of the number of compatibly communicating devices, as formulated by Metcalfe. The continuing growth in the number of subscribers to the public and private networks is the key element in maintaining that growth—as well as driving the eventual shifts away from desktop personal computers and toward a much wider range of devices.

The powerful impact of these observations can be seen when we add Moore's Law into the equation. Dr. Moore observed that the number of transistors that can be placed inexpensively on an integrated circuit doubles roughly every two years.

While Metcalfe's Law shows us that expanding the network increases its value, Moore's Law tells us that as manufacturing efficiency of the key elements steadily increases, the cost of those capabilities continually drops. When the observations of both Moore and Metcalf are considered together, the two observations point toward continued network expansion in the foreseeable future.

As we look forward to the next wave of technology that is still being commercialized and gaining in efficiency, it will become incrementally more feasible and economically worthwhile to network-enable even the simplest of devices and applications. This next wave of the Internet of Things (IoT)—as opposed to a communications path for individuals—is already being developed as more sophisticated industrial devices become network-enabled. The impact of that trend will also be explored in a later chapter.

Because of the number of machines using the communications networks, the total number of cellular subscribers in the U.S. and other countries begins to outnumber the total population. Modern communications have become so cheap, as demonstrated by the incessant texting of today's users, that the importance of the content has almost become secondary. The experience has been a boon for consumers, but it has saturated the cell-phone market and forced cellular companies to look elsewhere for revenue growth as the minutes of talk time per connection have flattened out.

Texting, interestingly enough, continues to climb at an annual rate of 74%, despite the fact that it's been called possibly one of the crudest forms of human-to-human communication since the pictograph. Still, even when combined with sales of additional consumer data plans, it remains challenging to sustain the kind of network growth required of the cellular companies.

However, the potential for becoming the dominant application for cellular networks appears to be in M2M communications. Moore's Law predicts the rate at which computer performance is increasing and also explains that connecting machines will steadily become cheaper and easier. Metcalfe's Law tells us that connecting millions of machines all around the world will add additional value to the cellular network and will continue to increase the demand for more processors.

Mike Fahrion, writer for Electronic Design, made a similar observation on May 3, 2012, regarding the future impact resulting from these two observations.

> "Looking backward, we can see how technology drove the shift from hunter to farmer, from farmer to factory worker, and from factory worker to knowledge worker. Looking forward, it's become clear that the convergence of Metcalfe's Law, Moore's Law, and the Internet of Things will produce another of these pivotal societal changes in just the next few decades."[1]

---

[1] http://electronicdesign.com/author/mike-fahrion

CHAPTER 6

Architectural Impact of Digital Wave

The major impact of the Second Digital Wave can be seen in three different areas.

The first impact of the Second Digital Wave is in the kind of data that is being created. The desktop personal computer was a tremendous boost to personal productivity when the user primarily remained at a desk and the nature of the data was delivered mostly in spreadsheets, charts, and graphs for business applications. However, as more content is generated by and prepared for mobile users and consumers, the nature of the information content changes from spreadsheets to more visual content. The Internet has also enabled more webcasts to be used as a method of dispersing information, and the amount of video content being processed is increasing.

IDC has estimated that by 2017 unstructured data will account for 79.2% of capacity shipped and 57.3% of revenue,[1] with file formats such as Microsoft Word, PowerPoint presentations, PDFs and non-textual content like images, graphics, video, and audio files. The percentage of unstructured data is continuing to grow. Only 10 percent of data being created today consists of spreadsheets and other formats that represent what until quite recently was considered core data for a business.

The target application for the development of new technologies has thus begun to shift from the infrastructure supporting a PC-oriented environment toward the development of networks necessary to process, develop, and store a wider range of data formats, as well as new tools that can act in response to this much data content.

To complicate the administration matters more, a modern storage array typically has dozens of other tunable parameters to improve the performance and thru put. Each of these settings has dozens of additional possible settings and priorities of providing efficient cloud services. The configuration matrix would thus require almost constant adjustments in priorities in order to maintain the highest level of efficiency and performance.

Due to the complexity presented by the number of performance options, the solution and administration of these services is now often the responsibility of software-based policy engines (i.e., software robots) to constantly direct the capabilities and service levels in order to maintain the highest levels of performance.

---

[1] http://www.idc.com/getdoc.jsp?containerId=247106

IBM and other companies define the challenges associated with processing, electronically transporting, and storing this much information as "big data." IBM's model consists of four separate challenges of "big data."

These challenges are separated into the issues associated with the velocity at which data needs to move about, the variety of possibilities of how that data is structured, the volume of digital data in the file, and the veracity or potential accuracy of the data.

On top of these computational challenges, an additional impact is the rapidly increasing amount of information being transmitted over mobile wireless networks. This is particularly challenging when it is considered that previously almost all of this data was exclusively transmitted over wired network communications channels.

Projecting that future rate of growth for wireless traffic is problematic. The potential for rapidly expanding the transfer of data over wireless networks is unquestionably a concern. At some point in the growth curve, that wireless communications channel is limited by the cost-effective expansion of the network. The ultimate resolution of that issue depends on the prioritization of available radio frequency bandwidth among competing applications. Resolution of this highly political decision of the allocation of communications resources eventually requires weighing choices relative to the cost, demand, and potential benefits.

An equally challenging non-technical issue that is easily overlooked is the ability of each cultural entity to absorb so much new knowledge. In many cases, this much exposure to new information can create increased levels of stress within the social fabric of communities, and we will expand on this topic in a separate chapter.

However, the fundamental question that remains is how much this easy access to so much information really benefits us individually as well as culturally and politically. While access to more information aids us in our daily business and personal life, such as using Wikipedia, "googgling" a person or event, or some other source to gain a quick understanding of a topic, are there any other benefits to having access to this much information? What do we truly gain? What do we sacrifice? Is there a long-term social benefit that will sustain all this effort?

From the point of view of increasing our knowledge base, Walter L. Warnick, in the Office of Scientific and Technical Information of the U.S. Department of Energy, discussed the importance of this broad data access to information as early as 2006. In a presentation that year at the annual meeting of the American Association for the Advancement of Science, his topic was "Increasing the Pace of Knowledge Diffusion to Increase the Pace of Science."

His opinion was that the development of science is related to the flow of information, as new methods, instruments, techniques, concepts, and data were created. These flows of information appeared to him to be diverse, complex, and initiated from practically all directions. He felt that "this incredible diffusion of knowledge is the very fabric of science. Given that the diffusion of knowledge is central to science, it behooves us to see if we can accelerate it. We note that diffusion takes time. Sometimes it takes a long time. Every diffusion process has a speed. Our thesis is that speeding up diffusion will accelerate the advancement of science..."

In order to accelerate this growth of knowledge, his group focused on a parameter commonly called the *contact rate*. Drawing from an example within the medical profession regarding the spread of diseases, the contact rate at which people come into contact with a person who has a particular contagious disease increases the rate at which that disease spreads. Dr. Warnick concluded that, in a similar manner, increasing the contact rate of new information would accelerate the diffusion of new information and in turn increase the rate of discovering new knowledge.

Dr. Warnick specifically addressed the challenge of increasing the flow of critical information:

"To do this we must reduce a huge gap in how the Internet works today…Even today, the Internet must still be considered new. Its emergence as a useful medium for communication really only dates from the first half of the 1990s, just a dozen years ago or so. It is a transformational technology still in its infancy…"

From slide 14 of Dr. Warnick's presentation:
"We stand on the rim of a new era of global discovery: Next steps

1. Expand content . . .

   • Increase searchable content within databases
   • Incorporate databases from all scientific communities
   • Beyond text—numeric data, audio, video, etc.

2. Enhance precision searching . . .

   • Integrate analytical tools
   • Improve sophistication and speed of relevancy ranking
   • Next-generation algorithms
   • Visualization techniques

3. Amplify computing power . . .

   • Deploy emerging technologies to enable extraction of ever increasing content
   • Increase computer power, storage capacity
   • Special architecture, grid technology"

In his concluding comments, Dr. Warnick noted that "To the extent that knowledge about new methods and concepts is spread more quickly, science itself will be accelerated…The calculus of innovation is really quite simple:

Knowledge drives innovation;

Innovation drives productivity;

Productivity drives our economic growth.

That's all there is to it."[2]

While there are clear advantages to be gained from new information and services, unfortunately it cannot be denied that the implementation and commercialization of some new information as well as the rate at which that information is being developed does put pressure on the social fabric of any society. In the next several chapters, we will look at some of the specific areas in which new technologies and capabilities are being developed and assimilated into societies.

---

[2]https://www.osti.gov/speeches/fy2006/aaas/

CHAPTER 7

Social Impact of the Digital Revolution

The previous chapter presented the advanced capabilities for mass communications between machines and individuals for monitoring and control purposes. What impact has the Digital Revolution had on mass communications among individuals?

According to Dr. Makimoto's "Digital Nomad," the concept was expected to provide liberation from:

- State monopolies that control telecom networks

- Prohibitive costs of computer power

- Geographic ties of work and school

- Political liberation from repressive regimes

Several of these items have been addressed to some extent, although the results vary from place to place. However, as technology and scientific knowledge increases, what impact will it have on societies and organizations that are not yet prepared for this accelerating rate of new ideas and options?

While potential conflict between government structures due to the difficulties of taxing and controlling the activities of these wandering professional workers can be resolved, the relationship between the Digital Revolution and a social revolution may be the more challenging aspect. This aspect of the Digital Revolution is outside of the scope of the technology itself, while the full extent of this impact is extensive and is still unfolding.

As frequently demonstrated in recent events, localizing public acts of defiance and disruption by controlling access to public communications services has become *de rigor*. The ability to organize—or contain—those events of social protest has been an effort with varying results to define new rules of engagement relative to the public communications networks between authorities and potential protesters.

For example, one reporter described the impact of the Digital Revolution on events in the Middle East in April 2012 in *Al Arabiya News*.[1]

---

[1] http://english.alarabiya.net/views/2012/04/06/205837.html

"Wielding mobile phones and computers, the young activists across the Middle East have altered the way the world approaches popular mobilization, social networks and Internet freedom. The Internet can be a transformational force for societies and individuals, allowing for organization on a mass scale and the free flow of information. However, we must remember that the Internet and social media are tools that do not bring change themselves, but act as facilitators in spreading the ideas. The seminal use of social media as vehicles for change in the Arab Spring uprisings exemplifies the power of web-based communication and makes a strong case for Internet freedom."

One common traditional response by authorities in the face of civil disobedience has been to limit the access to public communications networks. The ultimate reason the Internet in Syria became unstable during recent events in that area remains obscure. In some parts of that country even the most robust backup plans fell apart. In Damascus, activists sometimes use satellite phones for emergencies, minimizing their chances of being tracked by connecting for short amounts of time while driving aimlessly or calling from remote parts of town, far from their usual areas of operations.

In terms of high-visibility activities, the disclosure by Edward Snowden of the full extent to which the U.S. was monitoring digital traffic over public communications channels also continues to reverberate. Those activities related to recording and storing communications data had long been discussed within the technical community. However, confirmation of the practice, as well as the magnitude of the programs, has also led to acknowledgement by other countries that similar programs were also in place within their own boundaries. At the time that this is being written, debate also still continues on how much encryption capability and digital privacy should be available to private citizens.

On a much smaller scale, another pertinent example of the often first-response actions of government officials was a decision by San Francisco's Bay Area Rapid Transit (BART) officials. Faced with the prospect of large demonstrations that would use the subway system to deploy to various protest sites, BART officials decided to cut off underground cell phone service at some stations for several hours in order to tactically thwart the planned protests. However, in the rush to limit the impact of the protests, the city over looked the impact on business commuters at stations from downtown to the city's main airport as well as the fact that medical first responders also rely on incoming "911" emergency calls over those same networks.

The combination of these events along with many others demonstrates that government policies related to information access and usage of public networks are still evolving in response to rapidly changing technologies.

Officials in other countries have frequently been suspected of—or at least accused of—turning off the cell towers in the vicinity of large gatherings as a technique for disrupting and controlling large crowds. At a recent large demonstration in Russia, a completely different tactic for crowd control was also unveiled.[2] As usually happens in any large public protest, the line of contact between the protestors and the authorities is where the most threatening and dangerous activities take place. Shortly after midnight in this particular case, the cell phones began to ring for journalists, activists, and others in the vicinity of the conflict, as the activities between riot police and protesters continued to escalate. They had received a message. The English translation of that message was reported to have said, "Dear subscriber, you are registered as a participant in a mass disturbance." As we have seen in many cases, the first reaction to what is technically possible does not always become the pattern for future actions. There is still a learning process that takes place as new technologies get integrated into existing social structures.

In the case of the San Francisco transit authorities previously mentioned, the governing board considered the various arguments—along with suggestions from the Federal Communications Commission (FCC)—when later reviewing the event with the intent of considering the wider implications. In particular, the board wanted to consider a new policy outlining when and how the transit agency may shut down the antennas that provide cell phone service to its underground stations in response to any future events. The resulting policy approved by the San Francisco transit authorities permits the transit district to turn off cell phone service "only when it determines that there is strong evidence of imminent unlawful activity" threatening safety, property, or service. It requires the agency to limit the use of the tactic to areas and time periods where it is needed, and requires the decision to be made by the general manager or a specific designated person and process.

Specific examples of the kind of extraordinary circumstances that would warrant the shutoff of cell phone access were also cited, such as evidence of plans to use cell phones to detonate explosives, endanger BART passengers or employees, or facilitate plans to destroy district property or "substantially disrupt" transit service.

On a broader scale, the United Nation's Human Rights Committee affirmed that the protections guaranteed by the International Covenant on Civil and Political Rights (ICCPR) also applies to online communication. This announcement further established that bloggers should have the same protections as journalists, and also brought attention to the issue of how much personal information governments should access regarding individual citizens.[3]

It is to be expected that the social value of any new high-impact technologies may need some time to be fully integrated into the social and cultural fabric of any culture. The value of access to the digital communications networks is being acknowledged and often being codified as a fundamental social service.

However, it is not necessarily the arrival of any particular new technologies that is the ultimate issue. It is the accelerated rate at which multiple new technologies are arriving that is putting pressure on established social structures. This impact is effectively another aspect of the

---

[2]npr.org/sections/thetwo-way/2014/01/21/264537418/ukraine-tracks-protesters-through-cellphones-amid-clashes

[3]https://www.eff.org/deeplinks/2013/06/internet-and-surveillance-UN-makes-the-connection, see also: http://www.ohchr.org/Documents/HRBodies/HRCouncil/RegularSession/Session23/A.HRC.23.40_EN.pdf.

"contact rate" mentioned in a previous chapter. In some cases, ideas and information are being received faster than can be processed and integrated into some cultures.

In many cases, the new patterns of cause and effect are simply too complex for anyone to anticipate all of the ramifications in advance. In that sense, San Francisco's initial response to quell a disturbance followed by a subsequent after-action refinement of their future policy will likely become a familiar model that will be repeated at each new major advance in digital technologies.

In the following chapters, we will also consider some other potential future challenges.

CHAPTER 8

Other Unanticipated Consequences

One aspect of the unanticipated impact of digital communications is the ease with which large amounts of data can be transferred or transported. This of course has benefits as well as risks.

For example, the amount of data that may have been transferred to Wiki Leaks by U.S. Army Private Bradley Manning was hard to imagine. Accused of leaking classified U.S. information to Wiki Leaks in 2010, Private Manning was arrested in May 2010 after allegedly leaking more than 250,000 U.S. diplomatic cables, 400,000 U.S. Army reports, and an additional 90,000 documents about Afghanistan.

The sheer weight and bulk of transferring that much information would normally have been a challenge. However, that amount of data now seems relatively small compared to what was apparently in the possession of Edward Snowden, former CIA employee. Government officials have been unable to even determine how much information had been gathered.

We have also had the revelation of the amount of personal communications that is being monitored, stored, and often exchanged among most technically advanced countries. Public exposure of the fact that the programs extended to the monitoring of the personal cell phone of an ally leader of one of the most industrial countries was treated in the press as if it were more like "diplomatic rude behavior" rather than a major breach in the apparently naïve faith in public communications channels.

The U.S. National Security Agency's controversy resulted from the warrantless surveillance of private individuals within the United States during the collection of foreign intelligence by the National Security Administration (NSA). Nevertheless, the U.S. Senate recently renewed the warrantless wiretapping program within the U.S., which includes both voice and email messages.

The U.S. government claims sweeping authority under the Patriot Act to collect a record of every single phone call made by every single American "on an ongoing daily basis." Most of this issue was exposed in the broader discussion of what limits should be put on the NSA.

However, that is not the only ingredient in that bubbling pot. Two major security breaches occurred in 2014 in which hackers obtained personal information on over 22 million people. That data included addresses as well as specific information pertaining to that individual in two major breaches of U.S. government databases. That information reportedly included not only personal data, but also included information about other individuals listed as references in applications for security clearances.[1]

This activity was reported while other government groups are demanding that semiconductor designers include encryption backdoors in high-performance processors that would allow authorized data access by government agencies. Critics of the requirement maintain that backdoor access would create new vulnerabilities in those designs.[2]

At relatively the same time, the European Commission's Director of Fundamental Rights, Paul Nemitz, was drawing attention to the issue of personal privacy versus the public's right to know. His comments focused attention on the topic by declaring, "You have no right to see me naked!...A world in which everybody can see everybody naked is wrong."[3]

In truth, this particular comment was directed toward the issue of the personal right-to-be-forgotten and the personal right-to-be removed from search engines under certain circumstances. However, it did draw attention to the difference between what can be done compared to what we should have the ability to control.

Nevertheless, the unresolved challenge is maintaining the proper balance of these topics—and others that have not yet arisen—as the arrival of new technologies continues to accelerate.

Those reverberations relative to the control of and access to data have already manifested. Brazil, Canada, as well as some European countries have now requested that a less U.S.-centric "Internet" be established. Regardless of whether this particular modification comes about, it is also meaningful to note that a number of other countries have also acknowledged exchanging information collected from similar intelligence gathering programs.

Another example of the unanticipated impact of technology was the Stuxnet Internet virus. The Stuxnet virus is alleged to have been part of a U.S./Israeli operation that was initiated prior to 2010 to disrupt Iran's nuclear development program. A single USB drive was reported as the original source of the virus code that infiltrated the Internet in the 2005–2008 time frame. Originally inserted surreptitiously into the laptop of a single scientist, the intent of the original virus was to replicate itself within the computer system supporting the development of that specific nuclear energy program. The objective of the virus "payload" was to take control of the turbine used to generate power for the facility, and spin that turbine at an increased speed until either the facility exploded or—more likely—the facility and research would be shut down.

The unanticipated consequence was that the virus escaped onto the Internet, and the self-replicating software began to clog the performance of any computer or PC that unknowingly downloaded the code. Chevron's general manager of its earth sciences department reported to *The Wall Street Journal* that its network had been infected shortly after Stuxnet's discovery in July

---

[1]http://www.washingtonpost.com/blogs/federal-eye/wp/2015/07/09/hack-of-security-clearance-system-affected-21-5-million-people-federal-authorities-say/
[2]http://www.pcworld.com/article/2916912/lawmakers-criticize-fbis-request-for-encryption-back-doors.html
[3]http://www.theregister.co.uk/2014/11/05/right_to_be_forgotten_eu_panel/

2010 with similar symptoms. Mark Koelmel, general manager of the earth sciences department at Chevron, said, "I don't think the U.S. government even realized how far it had spread."[4]

However, in the case of Chevron and presumably other sites, even though the energy company had been infected, the Stuxnet software virus did achieve its task of withholding its payload after identifying a system that was not the original target. As a result, it caused no immediate damage to Chevron's systems and the company was able to remove the virus. Although Chevron's computer infrastructure wasn't adversely affected by Stuxnet's payload, the identification and removal of the malware does require action by all that were infected. This cost, while small, is significant when the total number of infected businesses is considered.

The International Space Station has also been hit with viruses that were carried on board unknowingly and injected into the system. In 2008, a virus that was designed to search for passwords was detected, and as recently as November 2013, another computer virus required that the entire control system be powered down and re-booted in order to remove the virus.[5]

The limit to which cell phone usage might also frequently be restricted due to its communications capabilities was addressed in a previous chapter. However, in many ways, the cell phone has already spawned other social issues. For example, one issue is that the ubiquitous usage of cell phones has permeated other aspects of almost every culture. Lawmakers in California have approved fines for all drivers using handheld cell phones, and also prohibit drivers under 18 from using hands-free phones. A third law also prohibits texting while driving.

In addition to the use of cell phones as a communications tool during periods of social unrest, other social challenges have risen as access by a broader group of individuals is often interpreted by some authorities as a loss of social control and a challenge to traditional behaviors. A village council in the eastern Indian state of Bihar has banned the use of mobile phones by women, saying the phones were "debasing the social atmosphere" by leading to elopements.[6] In addition to the ban, another village council has also imposed a fine of 10,000 rupees ($180) if a girl is caught using a mobile phone on the streets. One local person opposed to the fine said it was "disappointing" that the village council ignored the many advantages of mobile phones before placing a ban on them. "I want every girl to be given a mobile phone so that she could call up family members if she has a problem," he said.[7]

The ubiquitous usage of mobile communications has permeated almost all levels of social behavior. Historically, the physical presence of another individual in close proximity typically resulted in some level of acknowledgement or awareness of that person. However, with the advent of headphones and texting the "audience" that is occupying the highest level of attention may not be obvious or even physically present.

---

[4] http://blogs.wsj.com/cio/2012/11/08/stuxnet-infected-chevrons-it-network/
[5] http://www.theguardian.com/technology/2013/nov/12/international-space-station-virus-epidemics-malware
[6] http://timesofindia.indiatimes.com/india/UP-village-council-bans-jeans-mobile-phones-for-girls/articleshow/49043739.cms
[7] https://globalvoices.org/2012/12/08/women-banned-from-using-mobile-phones-in-indian-villages/

While this shift in social mores has typically been associated with the usual changes in behavior expectations between younger and older generations, other technology advances such as Google Glass are also crossing established boundaries. This style of eyewear expands the features of the Internet and other digital technologies in a manner that many find too intrusive. For example, the ability to combine facial recognition features along with Internet access to Facebook allows the wearer to scan a crowd in search of a particular individual. Other gesture-control capabilities allow the wearer to unobtrusively record a digital image of the immediate scene. Some public social environments have banned the devices for the moment, and the company has recently withdrawn the glasses from the market until a less obtrusive version can be released. Other groups are calling for a new set of social behavior guidelines to be developed, and Google has also recently suggested a code-of-conduct for users. Nevertheless, it is clear that there is a market for the features of Google Glass. Future technology development will likely make the devices less obvious, and potential widespread public acceptance will eventually lead to broader usage similar to that of allowing in-flight cell phone usage on commercial aircraft.

The technology issue is not about being unable to decide the proper usage of technology consistent with the mores and expectations of the particular society. As seen by the San Francisco experiences from shutting down cellular services described previously the typical first reaction of any organization is to focus on containing any anticipated disruptions. It is only through experience and in the after-action review that the full extent of a new technology—along with the unanticipated consequences—can be acknowledged and a judgment made that reflects the balance among the various goals and values of the organization.

At the heart of that issue is a topic that continues to be debated. The issue is the ability to correctly predict and anticipate the actual rate at which new information and technological capabilities can be absorbed into various societies. Is technology being advanced at a rate that is consistent with the social and intellectual advances of the society—or is technology and knowledge now arriving at an accelerated rate that may be progressing faster than the historic rate of human evolution?

As we enter the Third Digital Wave of robotics and artificial intelligence anticipated by Dr. Makimoto, the key concern is whether we will trigger a social evolution on the scale of the Industrial Revolution—or perhaps even an irreversible evolutionary shift on a much higher scale of disruption.

The book *Digital Nomad* contained an additional comment regarding the impact of this broader exposure to other cultures that can result from this expanded capability of travel and communications.

"Almost certainly the dominant role of governments is set to diminish as national boundaries have less and less relevance to people and the pressures of global markets largely decide their policies for them...As the influence of government declines, and people's ties to geographic regions weaken, people will probably give their primary social allegiance to a group rather than to their country or origin. They might give it, for instance, to a company, to a sect, or to an interest group.

Who can tell? When social change meets new technological opportunities to release long-suppressed human instincts, the result could be the biggest revolution in human behavior for 10,000 years—since humans relinquished the life of nomadic hunter-gatherers and settled down to farm. Could ten millennia of settled existence turn out to have been a temporary aberration?"

In the light of recent disclosures to the amount of monitoring of digital traffic that is being conducted by various countries, perhaps it is worth pondering how various political structures and government agencies might react to any sense of weakening influence.

CHAPTER 9

Robotics: The Third Digital Wave

The eventual integration of the Second Digital Wave's consumer-level mobile communications into the fabric of society is not the end of the Digital Revolution; it corresponds with the beginning of its next phase.

The following chart from Dr. Makimoto, previously shown in an earlier chapter, has been updated to represent the transition from PCs to digital consumer and network applications, and eventually onward to the rising wave of robotics as the target application for the development of new semiconductor technologies.

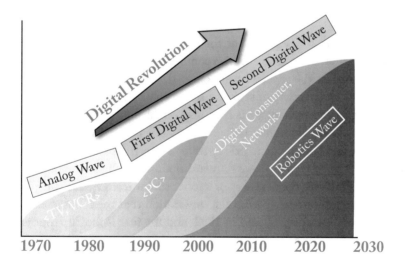

Figure 9.1: The Digital Revolution.

The next chart had also been prepared to describe the declining influence of technologies that derive their value primarily from the increased number of transistors per area, and the anticipated additional value of a new type of "Cleverness Driven Device."[1]

---

[1]http://www.sony.net/Products/SCHP/cx_news_archives/img/pdf/vol_32/sideiview.pdf

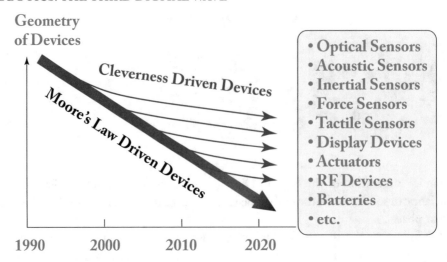

Figure 9.2: Geometry of devices.

The distinguishing characteristic of the group of products on this chart is the additional value that can be created by the cost-effective combination of unique performance features similar to those later included in the separate technology development described in previous chapters, and in particular, the wide variety of sensors used in robotics.

Other scientists and technologists have also established anticipated milestones for the development of technologies associated with robotics. Dr. Hiroaki Kitano of the Sony Computer Science Laboratory and Kitano Symbiotic Systems Project, Japan Science and Technology Corporation, first proposed "RoboCup" as a standard against which progress could be measured. The RoboCup concept measures the dexterity of a robot soccer team against the physical dexterity of the human world champion team—with the expectation that the robotic team will be comparable by 2050. This concept was proposed in 1992, and the first RoboCup tournament was held in Nagoya in 1997. The progress from year to year provides an opportunity to measure the growth in performance and dexterity of the participants.

Building on that concept, Dr. Hans Moravec of Carnegie Mellon University has predicted the likely rate at which advances in semiconductor process technologies will have to be achieved in order for robots to have that level of information-processing capabilities.[2]

As seen in the chart above, Dr. Moravec suggests that robots can eventually evolve to a level of equivalent processing power in terms of millions of instructions per second, and he has predicted the likely rate at which advances in semiconductor process technologies will have to be

---

[2]ibid.

Figure 9.3: Robots competing in 2012.

achieved in order for robots to reach a level of information-processing capabilities equivalent to the human brain. His prediction is the year 2040.[3]

The basis for many of these extrapolations is linked back to concepts presented by Dr. Moravec in his 1997 paper titled "When will computer hardware match the human brain?" In that paper, Dr. Moravec argued that the function of voice recognition and language translation of a switchboard operator, which includes the human nervous system and the recall of information stored in the human brain, requires about $1/30^{th}$ of the total computational capabilities of the human brain. The reference point is approximately 3.3% of human intellectual capability for that level of intellectual activity.

However, while we still have that image of switchboard operators in mind, let's look quickly at "Amelia." Her current level of performance is an example of the current level of development.

Amelia[4] is not a mechanical robot. "She" is a software program—presented as the "First Cognitive Agent" fro IPsoft. She appears on a screen as an image of a twenty-something young woman. Her role is to establish a new type of computer (artificial) intelligence that can absorb, deconstruct, and use information like a human. In a recent demonstration, Amelia was given the entire text of a complex technical manual. Within seconds, Amelia was ready to respond to questions on any topic in the manual and reply in 20 different languages with an accurate and useful answer. In the event that the question could not be found in the material that Amelia had recorded, she routes you to a human helper. Amelia then listens in on that discussion, deconstructs how the problem was solved, and includes that information in her own memory bank. The

[3]http://www.transhumanist.com/volume1/moravec.htm
[4]http://www.ipsoft.com/what-we-do/amelia/

Processing Power (MIPS)

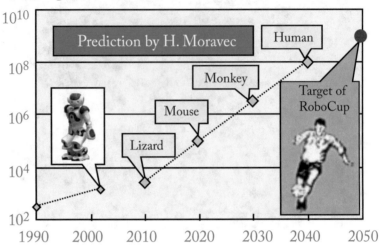

Figure 9.4: Evolution of robotic intelligence.

ultimate objective for Amelia is to allow her human supervisor to be able to train her by speaking to her the same way you would speak to any other new employee in your office.

While demonstrating that level of verbal skill is extremely impressive, what practical and commercial applications could be developed that would encourage such an endeavor?

One obvious area of value is the growth of industrial robots that perform repetitive tasks, and we have all seen images of industrial robots moving boxes and materials from one part of a warehouse to another. The International Federation of Robotics reports that 179,000 industrial robots were sold worldwide in 2013 as an all-time high and 12% more than in the previous year. Similar growth is also expected in 2014.[5] Another emerging target market is in the smaller category of service robots for personal and domestic use.[6]

The robot garnering the most attention at the moment is "Pepper," who is being trained in Japan as a caregiver. Pepper has been taught to focus on voice tones, facial expressions, and emotional content to give him the ability to learn about and understand human interaction. The 4-foot-tall robot can also sing, dance, and tell jokes. The first 1,000 models were offered for sale in Japan in June 2015 for a price of $1,600 and a monthly fee of $200. The robots all sold within the first minute after they went on sale.

Pepper also relies heavily on a communications infrastructure. Each robot shares its experiences with other identical models by up loading all of their activities to a cloud computer system

---

[5]http://www.ifr.org/news/ifr-press-release/global-robotics-industry-record-beats-record-621/
[6]http://www.ifr.org/service-robots/statistics/

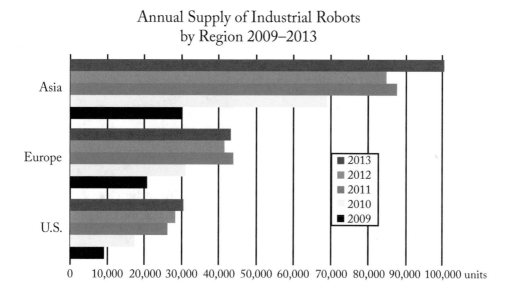

Figure 9.5: Supply of industrial robots.

for other artificial intelligence robots of the Pepper design to analyze and use. That shared pool of information and experiences is expected to further accelerate the development of the emotional capabilities of this class of robots. Pepper is expected to move into business applications as a store greeter, and later as a companion and helper for the elderly.

The goal of these projects is to not only take over many of the tedious and repetitive tasks in our daily lives, but to be integrated into our human society. However, the question that ultimately still remains is the cultural impact on human societies.

In the case of Pepper, it is specifically being developed to provide support to an aging community where shortages of caregivers in the future may be a concern.

As Amelia's designer has commented, "It's not a guarantee for the future, but I think this will be another shift and transformation of people working in a different context."

In the broader case of the general work environment, the assumption among many futurists and technologists is that these new technologies will create as many jobs as they eliminate. However, it is also worth noting that there is no reason to expect the new employment opportunities will necessarily be at the same skill level or in the same geographic location—or even in the same country—as the jobs that are eliminated.

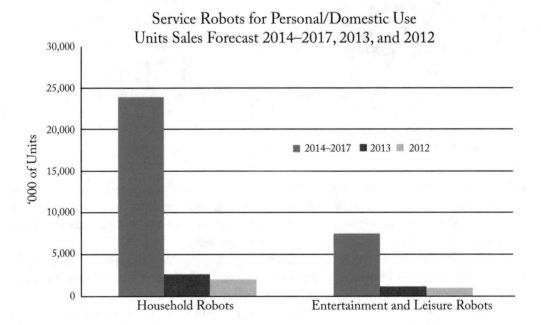

Figure 9.6: Service robots.

CHAPTER 10

Era of Cognitive Systems

In an earlier chapter, we discussed Metcalfe's Law that the value of a network is increased as a function of the number of active members on that network, and we also identified the network being assembled to support the network of robotic caregivers being assembled around Pepper. The salient feature of this information network was the regular gathering of information, and the group-wide sharing of any new ideas or suggestions.

In 2012, Dr. John E. Kelly III, IBM Senior Vice President and Director of IBM research, described a similar impact that would occur in the area of computer research. In addition to the increased capabilities relative to human intelligence, he envisioned that an increase in human skill sets would also likely come from the next level of computer development. Below is a lengthy but extremely applicable presentation by Dr. Kelly. The concept is very similar to the discussion presented in an earlier chapter relative to increasing the value of networks.[1]

"Today, we are at the dawn of another epochal shift in the evolution of technology. At IBM Research, we call it the era of cognitive systems...

How do we define the era of cognitive systems? It helps to compare it to what came before. The tabulating era began in the 19th century and continued until the 1940s. Those mechanical devices were used to organize data and make calculations that were useful in everything from conducting a national population census to tracking the performance of a company's sales force. The programmable computing era emerged in the 1940s when scientists built the first electronic programmable computers. Successive generations of computing technology enabled everything from space exploration to the Internet.

Cognitive systems are fundamentally different. Traditional computers, which are still based on the blueprint that mathematician John von Neumann laid out in the 1940s, are programmed by humans to perform specific tasks. Cognitive systems are capable of learning from their interactions with data and humans—essentially continuously reprogramming themselves. Traditional computers are designed to calculate rapidly. Cognitive systems are built to analyze information and draw insights from it. Traditional computers are organized around microprocessors. With cognitive systems, it's about the data and drawing insights from it through analytics.

---

[1]http://asmarterplanet.com/blog/2012/05/welcome-to-the-era-of-cognitive-systems.html

Because of these changes, the machines of the future will do much more than compute. They will be able to sense, learn and better predict the consequences of actions. In the years ahead, machines will cull insights from the vast amounts of information being gathered to help us learn how the world really works, and make sense of all of that complexity, and provide trusted advice to humans—whether heads of state or individuals trying to manage their careers or finances. Computing intelligence will become ubiquitous and pervasive. Increasingly, computers will offer advice, rather than waiting for commands.

This new era of technologies is essential to fulfilling IBM's goal of using technology to help create a smarter planet—to make the world work better.

We need cognitive systems because recent developments in business, society and technology require new capabilities. The emergence of social networking, sensor networks and huge storehouses of business information create a seeming overabundance of information that some call Big Data. Systems are being asked to find patterns and draw conclusions, often in near real time, from huge quantities of information—and in situations where precise answers are hard to find. At the same time, in fields ranging from retailing to healthcare to government, the individual increasingly stands at the center. People are newly empowered with information about how the world works and able to express themselves in powerful new ways. They're also becoming increasingly decipherable via the cloud of data that surrounds them and the digital exhaust they leave behind wherever they go. Through data analytics, we can know each other much better.

In addition, we need cognitive systems because some of the fundamental building blocks of traditional computing are crumbling. For instance, consider what's happening with silicon-based integrated circuits. Today's microchips are the electronic brains in everything from refrigerators to space ships. However, because of the laws of physics, we are no longer achieving improvements in the performance of microchips that we are accustomed to, and need, using traditional methods. We need to invent new materials and new chip architectures…

We're on the leading edge of a technology transformation that promises to utterly transform business and society once again. But even though IBM has a broad portfolio of technologies and expertise, no single company can handle this sort of thing alone. We look to our clients, university researchers, students, government policy makers, industry partners and entrepreneurs to take this journey with us. Welcome, all, to the era of cognitive systems."

In order to raise the level of performance of the robotic devices, the "brains" and "nervous system" of this robot would still has to be derived from the digital computing system. Bernard

Meyerson, Vice President of Innovation for IBM, published a paper in 2013 titled: "What's Next in Tech? Computers that 'Sense' Experiences." Here is IBM's comment on the rate of growth.[2]

"Today's computers calculate, automate processes and organize data. A new class of machines, however, is starting to solve problems by invoking logic more akin to how humans tackle challenges. To make this process efficient, such systems must also then experience the world much the way people do.

Over the next five years, computers will start to use the five senses of sight, sound, smell, feel, and taste to identify and then solve problems. These senses—how humans perceive their environment—will give computers the context they need to analyze problems in novel ways, helping make the world a more sustainable, productive, and enjoyable place.

In the process, these new systems will propel the world further into the age of cognitive computing, where systems learn from interacting with their environment, and help people make smarter decisions using that learning."

Dr. Meyerson addresses each of our five senses in turn and describes the potential capabilities of future computer designs. In duplicating our human sense of sight, he anticipates that:

"Computers will begin to automatically make sense of the reams of videos, photos, and drawings being created today. Rather than waiting for people to tag their visual creations, computers will learn this function by being taught what to look for by seeing previous examples, and then recognizing those patterns of interest to the user through that learning.

Computer systems will use "brain-like" capabilities to extract key information, understand its context, and determine the "meaning" of the content in a given image. Consumers and companies will be able to interact with such systems, asking questions requiring the system to potentially add further understanding of what they are seeing. That will make it possible to pull new insights out of digital visual data, whether it's examining medical MRIs to help diagnose a health condition or spotting patterns in home photo albums to organize them automatically."

Our sense of hearing sounds is also fundamental to our human experiences and understanding of the world. IBM anticipated that in five years time (i.e., by 2018),

"…People will control and personalize their experience at the events they attend through the use of multiple announcing systems. These systems will follow, interpret, and describe events on the field or at a show by understanding a person's focus by using local sensors, including cameras and microphones. At a soccer game, for instance,

[2]http://asmarterplanet.com/blog/2012/12/the-ibm-5-in-5-our-2012-forecast-of-inventions-that-will-change-the-world-within-five-years.html

a fan sitting in the stands watching a goalie could receive live commentary about that player's history and current performance. Sensors will observe the spectator's gaze and body movements to tailor the commentary to match an individual's interests, which will be delivered directly to him or her using directional audio, technology that can project a beam of sound so narrow that only one person can hear it."

A sense of smell is also a capability that can be duplicated. Mechanical systems already available today can sense environmental conditions such as temperature, humidity, airflow, and specific particles that may be present in the air.

IBM is already working with health care organizations to expand those capabilities by equipping patient rooms with sensors that monitor airborne information too faint to be detected by humans.

"Hundreds of sensors will sniff for cleanliness, detecting and identifying the chemical compounds found in the rooms and even on the hands of patients and staff to pinpoint, for instance, whether a patient's room has been cleaned.

Sensors will also sniff out potential diseases. By using devices that can analyze biomarkers and thousands of molecules in a patient's breath, doctors will be able to diagnose and monitor many ailments such as liver and kidney disorders, asthma, and diabetes by smelling whether concentration levels are normal or not."

Transferring a sense of touch is also possible—and already available to a limited degree. Technologies that use the variable frequency patterns of vibrations associated with different physical situations are already built into mobile devices and gaming devices to simulate the sensation of driving over a rough surface or the jolt of a car collision.

"Now, researchers are experimenting with using these technologies in other industries, making it possible for online merchants, for instance, to let customers touch merchandise before they purchase. Using the vibration motion of the phone, the texture of a piece of clothing can be simulated when the shopper brushes his or her finger over the item on the screen. Each type of fabric will have its own unique vibration pattern, with silk having a softer vibration, for instance, and linen having a stronger one."

Techniques are also under development that will enable a sense of taste for computers. Creating this sense of taste within a computer involves breaking down the ingredients to their molecular level. Once these elements are established, the machine can then blend the chemistry of the compounds with the psychology behind the flavors and smells that humans like. The next step for the computer is to analyze a database of millions of recipes in order to generate tasty and novel combinations. That system could be further refined to "cater to diners' personal preferences and dietary restrictions as well as constraints such as what local food is currently in season…"

"These five predictions show how cognitive technologies can improve our lives, and they're windows into a much bigger landscape—the coming era of cognitive systems.

The world is tremendously complex. We face challenges in deciphering everything from the science governing tiny bits of matter to the functioning of the human body to the way cities operate to how weather systems develop. Gradually, over time, computers have helped us understand better how the world works. But, today, a convergence of new technologies is making it possible for people to comprehend things much more deeply than ever before, and, as a result, to make better decisions."

"...IBM Research is taking the lead in producing some of the scientific advances that will enable the big shift to cognitive computing. A team at our lab in San Jose, Calif., for instance, is designing a chip that's based on the architecture of the brain that could become the brains of the railroad robot. The goal is to create a system that analyzes complex data from multiple senses at once, but also dynamically rewires itself as it interacts with its environment—all the while rivaling the brain's compact size and low energy usage. Our lead researcher on the project, Dharmendra Modha, envisions being able to package the computational power of a human brain in a container the size of a shoebox."

"...But the point isn't to replicate human brains. We humans are no slouches when it comes to procreation. And this isn't about replacing human thinking with machine thinking. Once again; not necessary. Rather, in the era of cognitive systems, humans and machines will collaborate to produce better results—each bringing their own superior skills to the partnership. The machines will be more rational and analytic. We'll provide the judgment, empathy, moral compass and creativity."

"Indeed, in my view, cognitive systems will help us overcome the 'bandwidth' limits of the individual human."

"–Limits to our ability to deal with complexity. We have difficulty processing large amounts of information that comes at us rapidly. We also have problems understanding the interactions of the elements of large systems—such as all of the moving parts in the global economy. With cognitive computing, we will be able to harvest insights from huge quantities of data, understand complex situations, make accurate predictions about the future, and anticipate the unintended consequences of actions."

"–Limits to our expertise. This is especially important when we're trying to address problems that cut across intellectual and industrial domains. With the help of cognitive systems, we will be able to see the big picture and make better decisions. These systems can learn and tell us things we didn't even ask for."

"–Limits to our objectivity. We all possess biases based on our personal experiences, our egos and emotions, and our intuition about what works and what doesn't. Cognitive systems can help remove our blinders and make it possible for us to have clearer understandings of the situations we're in."

"–Limits to our senses. We can only take in and make sense of so much stuff. With cognitive systems, computer sensors teamed with analytics engines will vastly extend our ability to gather and process sense-based information."

As computing devices and robots gain a wider range of human-like characteristics, the question that is more frequently considered is how are humans going to interface with and treat these new machines that share so many common features with mankind.

CHAPTER 11

The Uncanny Valley

At this point, it seems technically likely that robots will eventually acquire at least some set of skills and responses that would appear similar to those of humans. As robots acquire more skills and more familiar behavior, how might we treat these machines with expanded human-like capabilities? Is there some point at which we might feel obligated to treat them as more than simple machines?

There are two dominant scenarios that push beyond our current perception of computers and machines. One scenario is represented by the ability of a robot to perform functions normally associated with humans. It is generally assumed that these future robotic designs, such as Pepper and Amelia, can continue to gain more human-like dialog and behaviors. That possibility also assumes a continuum of transferring more human-like capabilities and attributes as specialized robots become more integrated into the general spectrum of human society.

The other scenario identified in a previous chapter suggests an accelerated advancement of human intellect and capabilities by way of physical attachments to the brain or technologically assisted replacements and enhancements of limbs or senses to the point that humanity absorbs so many attributes of machines that the distinction between human and machine becomes blurred.

From a practical point of view, the technologies necessary to enable either scenario are similar enough that it is easy to assume advances along both tracks will occur almost continuously and simultaneously—such as mechanical devices controlled by existing nerves or the ability to restore a sense of touch, vision, or hearing by way of prosthetic devices.

In some cases, the technology advances do not necessarily overlap. It is therefore still useful to separate the two technology tracks in order to measure the current level of technological development.

For example, in terms of computing devices performing human-like functions on the level of diagnostic capabilities, computer-driven devices are replacing humans in the initial interview and triage of new patients. Successful diagnosis of a patient's symptoms and the determination of the next level of medical attention is often more accurately performed by computer-driven logic. This success rate has been demonstrated over time where medical services are scarce and computers have served as the first-level contact with patients in remote locations.

At the same time, while an advanced machine design may excel in logic and diagnostics, the ability to perform surgery still favors a human practitioner. The physical aspects of surgical services have been remotely provided for some procedures through the Da Vinci Surgical System. However, even if the surgeon's knowledge base could be duplicated, combining those physical skills in a robot with the knowledge and sense of touch of a human does not seem to be on the horizon.

It's not too surprising that complex human physical activities would lag behind the expansion of mental capabilities when comparing a human to a programmable device. The expectation of a robot being able to play soccer at a professional level by 2040 gives an idea of when human-level dexterity may be possible for robots.

In the meantime, there are other physical activities that do not require that same level of agility. In those areas of physical prowess, robots are already beginning to demonstrate a growing repertory of skills.

DARPA's Robotic Challenge has been an excellent opportunity to gauge the progress. In DARPA's Robotic Challenge completed in December 2013, some teams entered the competition with machines they had constructed while other teams made use of Atlas—a robot manufactured by another Google-owned business, Boston Dynamics—and the competitors each controlled the machine with their own software. Team Schaft, a robot developed by a Japanese start-up company and now also acquired by Google, completely dominated that competition. More than 100 teams originally applied to take part, and they were reduced to 17 challengers ahead of the final event.

The final set of challenges tested the robots' ability to accomplish eight tasks normally associated with emergency situations. The tasks included climbing a ladder, driving a car, and maneuvering over difficult terrain. In order to encourage development of more adept robots the agency challenged contestants to complete a series of tasks, with a time limit of 30 min for each task:

- Drive a utility vehicle along a course

- Climb an 8-ft-high (2.4 m) ladder

- Remove debris blocking a doorway

- Pull open a lever-handled door

- Cross a course that featured ramps, steps, and movable blocks

- Cut a triangular shape in a wall using a cordless drill

- Close three air valves, each controlled by a different-sized wheel or lever

- Unreel a hose and then screw its nozzle into a wall connector

Team Schaft's machine was the only one to complete all eight tasks and outscored its rivals by a wide margin.[1] In 2014, the updated version of the Schaft robot (or S-1) won four of the eight tasks.

Another robot, not specifically designed for the DARPA tests, is already on assignment. Samsung's SGR-A1 is a military robot sentry already deployed to the demilitarized zone (DMZ)

---

[1] http://www.dvice.com/2013-12-23/team-schaft-takes-top-honors-darpa-robotics-challenge

Figure 11.1: Robot 1.

at the border of North and South Korea. The robot, which was developed by a South Korean university, uses "twin optical and infrared sensors to identify targets from 2.5 miles (4 km) in daylight and around half that distance at night." It is capable of tracking multiple moving targets using IR and visible light cameras, and is under the control of a human operator. The Intelligent Surveillance and Guard Robot can "identify and shoot a target automatically from over two miles (3.2 km) away." It is also equipped with communication equipment (a microphone and speakers), "so that passwords can be exchanged with human troops." If the person gives the wrong password, the robot can "sound an alarm or fire at the target using either rubber bullets or a swivel-mounted Daewoo K-3 light machine gun." During their deployment to Iraq, South Korea's soldiers were also reported to have used a mobile version of the SGR-A1 robot as sentries to guard home bases.[2]

---

[2]http://en.wikipedia.org/wiki/Samsung_SGR-A1

Figure 11.2: Robot 2.

As an example of another functioning robot, here is the Wildcat from Boston Dynamics.[3] Designed as an all-terrain robotic "packhorse," the Wildcat is capable of running on all types of terrain and has reached speeds of about 16 mph on flat terrain.

Figure 11.3: Wildcat All-terrain Robot.

---

[3]http://www.extremetech.com/extreme/168008-meet-darpas-wildcat-a-free-running-quadruped-robot-that-will-soon-reach-50-mph-over-rough-terrain

When comparing these various robots to the soccer-playing robots, it is obvious that they have different physical appearances that may impact the initial responses of humans to these devices.

Figure 11.4: Robot 4.

One concept for measuring the human acceptance and reaction to robots can be seen in the work of Dr. Masahiro Mori of the Tokyo Institute of Technology. Dr. Mori was one of Japan's early and leading scientists developing robots. In 1974, Mori published *The Buddha in the Robot: A Robot Engineer's Thoughts on Science and Religion*, in which he discussed the metaphysical implications of robotics. In 1988, Dr. Mori founded the first nation-wide robot-building competition in Japan and has widely promoted robot competitions in the years since then. Dr. Mori is currently president of the Mukta Research Institute, which he founded in Tokyo in order to promote his views on religion and robots. The institute also provides consultation on the use of automation and robotics in industry.

His original hypothesis stated that a human observer's emotional response to robots becomes increasingly positive and empathic as the appearance of a robot is made to resemble a human. However, as the robot's appearance continues along that spectrum to the point of becoming less distinguishable from a human to the point of creating confusion to the viewer, a person's emo-

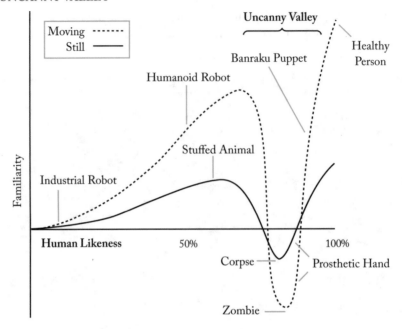

Figure 11.5: The Uncanny Valley.

tional response quickly becomes one of strong revulsion. However, if the form is clearly known to be a robot, as the robot's appearance becomes less distinguishable from that of a human being, the emotional response becomes positive again and actually approaches human-to-human empathy levels.

Dr. Mori called the area representing a negative response to the human-like appearance and motions between a barely human and a fully human-like entity as "the uncanny valley." The "uncanny valley" model postulates that the uncertain identity of an almost-human-looking robot creates a wariness and discomfort in a human being, and will therefore fail to evoke the empathetic response required for productive human-robot interaction.[4]

The company Rethink Robotics was particularly sensitive to this aspect in their construction of Baxter and Sawyer, both robots intended to work in close proximity with humans. "Designers had to negotiate the 'Uncanny Valley,' a term that refers to the discomfort people experience when they encounter a robot that looks too 'human.' Because Baxter is taught its movements directly by a person manipulating its arms, …[the designers] wanted the robot to convey an empathetic quality that would make Baxter's human coworkers comfortable."[5]

---

[4]https://en.wikipedia.org/wiki/Uncanny_valley
[5]http://www.farmpd.com/product-development-projectst/rethink-robotics

As the technologies progress and robots begin to gain more intellectual capabilities, the question arises again of how do humans interface with and treat these new machines—and should they also gain a higher social or legal standing than traditional machines as the robots gain so many human-like capabilities?

For example, the ability of humans to draw a logical conclusion from certain data is associated with becoming self-sufficient and making decisions in one's own best interests. How much self-awareness and ability to maintain its own existence should be given to robots?

In the late 18th century, Immanuel Kant maintained only beings that are self-conscious should have moral standing. At the time that he expressed this view, it was believed that all—and only—human beings are self-conscious.

The traditional test for self-consciousness is to demonstrate responses consistent with using a mirror to inspect marks temporarily placed on the body. So far, only adult humans and great apes have consistently demonstrated those capabilities, while very young children lack self-consciousness. However, some also suggest that other species such as monkeys, dolphins, lesser apes, and elephants might also qualify if the test was less reliant on touching a marked part of the body with a hand.[6]

Basing a higher moral standing on just this single characteristic therefore might not justify our opinion of having achieved a higher level of evolution. According to Kant's criteria, we might be entitled to grant some animals as well as presumably intelligent ethical machines with a higher regard than lower level animals—even if we decide they should not have quite the full moral status of human beings. We can force them to do things to serve our objectives; however at the same time we should not mistreat them.

That sense of self-awareness for a robot is also no longer a rhetorical question. In a test of three robots at New York's Rensselaer Institute, a version of a riddle that only a person is usually able to pass was used to see if a robot is able to distinguish itself as a separate entity.[7]

In this test, all three robots were programmed to believe that two of them had been silenced. In order to claim "self-awareness," the remaining robot would have to understand the rules, recognize its own voice, and be aware of the fact that it was a separate entity from the other robots.

When all three robots were simultaneously asked which of them could still speak, only one was able to say, "I don't know" out loud. However, upon hearing its own voice, that robot realized it was the only one that could speak and immediately changed its answer.

This topic of computers gaining a sense of self-awareness is not as esoteric as it first appears, since we already have computers with the ability to reconfigure and re-assign resources on the fly in order to increase the overall efficiency of the system. That sense of awareness is particularly important when considering the ability of programmable machines to very quickly acquire enough autonomy and intelligence necessary to become dangerous.

---

[6] http://www.pnas.org/content/98/10/5937.full
[7] http://www.businessinsider.com/this-robot-passed-a-self-awareness-test-that-only-humans-could-handle-until-now-2015-7

CHAPTER 12

The Human Interface to Advanced Robotics

Vernor Vinge and other futurists have suggested that computers might—unintentionally or intentionally—have the ability to suddenly and unpredictably become thousands or possibly millions of times more intelligent than humans over just a few years. Stephen Hawking commented in a recent article:[1]

> "One can imagine such technology … out-manipulating human leaders and developing weapons we cannot even understand."

> "If a superior alien civilization sent us a message saying, 'We'll arrive in a few decades,' would we just reply, 'OK, call us when you get here—we'll leave the lights on'? Probably not, but this is more or less what is happening with AI."

The award-winning theoretical physicist and mathematician concluded that inventors and researchers currently pushing the artificial intelligence boundaries aren't considering the potentially massive negative outcomes if a short period of very rapid acceleration of knowledge is achieved.

Hawking says the benefits could be huge for artificial intelligence, but his concern is that not enough care is given to how these inventions could be controlled in the future. His article continued:

> "Artificial-intelligence (AI) research is now progressing rapidly. Recent landmarks such as self-driving cars, a computer winning at Jeopardy and the digital personal assistants Siri, Google Now and Cortana are merely symptoms of an AI arms race fuelled by unprecedented investments and building on an increasingly mature theoretical foundation. Such achievements will probably pale against what the coming decades will bring…"

> "The potential benefits are huge; everything that civilization has to offer is a product of human intelligence; we cannot predict what we might achieve when this intelligence is magnified by the tools that AI may provide, but the eradication of war, disease, and poverty would be high on anyone's list. Success in creating AI would be the biggest event in human history."

---

[1]http://www.independent.co.uk/news/science/stephen-hawking-transcence-looks-at-the-implications-of-artificial-intelligence-but-are-we-taking-9313474.html

"Unfortunately, it might also be the last, unless we learn how to avoid the risks… Looking further ahead, there are no fundamental limits to what can be achieved… One can imagine such technology outsmarting financial markets, out-inventing human researchers, out-manipulating human leaders, and developing weapons we cannot even understand. Whereas the short-term impact of AI depends on who controls it, the long-term impact depends on whether it can be controlled at all."

"So, facing possible futures of incalculable benefits and risks, the experts are surely doing everything possible to ensure the best outcome, right? Wrong.… Although we are facing potentially the best or worst thing to happen to humanity in history, little serious research is devoted to these issues…"[2]

Not only do we have to deal with these machines that will eventually share so many common features with mankind—we also have to consider if some of these machines in fact might eventually acquire a higher legal standing than traditional machines.

In the meantime, advances in AI (artificial intelligence) and the creation of more intelligent robotics are two of the most easily recognized technologies that are fundamental for accelerating the transition toward other new technologies.

In fact, some machines have already acquired various forms of semi-autonomy. Examples include the ability to locate their own power sources, to choose targets to attack with weapons, and the capability to optimize computer configurations on an on going "live" basis to achieve more efficiency. In addition, some computer viruses have already achieved a level of awareness to evade elimination that is considered to be at a comparable at least a "cockroach level of intelligence."[3]

The Association for the Advancement of Artificial Intelligence commissioned a study to examine this issue, and highlighted particular programs like the Language Acquisition Device, which can emulate human interaction.

In their study, they concluded that some level of support for the design of friendly artificial intelligence already exists, and that future advances in the development of AI would also likely include an effort to make AI intrinsically friendly and humane.

Even prior to these guidelines for dealing with robots, researchers in the United Kingdom had been exploring "machine intelligence" for up to ten years prior to the recognition of the field of artificial intelligence research in the mid 1950s. The informal group of British cybernetics and electronics researchers that included Alan Turing had already had frequent discussions among the members of the Ratio Club.

Turing, in particular, had been considering the concept of machine intelligence since at least 1941 and had made one of the earliest known mentions of "computer intelligence" as early a 1947. In Turing's report, "Intelligent Machinery," he investigated the question of "whether or not it is possible for machinery to show intelligent behavior."

---

[2]Ibid.
[3]https://en.wikipedia.org/wiki/Ethics_of_artificial_intelligence

As part of that investigation, he proposed a format in that report which is similar to his later tests:

> "It is not difficult to devise a paper machine, which will play a not very bad game of chess. Now get three men as subjects for the experiment. A, B and C. A and C are to be rather poor chess players, B is the operator who works the paper machine.... Two rooms are used with some arrangement for communicating moves, and a game is played between C and either A or the paper machine. C may find it quite difficult to tell which he is playing."

"Computing Machinery and Intelligence" was published in 1950 in *Mind*, a British peer-reviewed academic journal published by Oxford University Press. This paper by Turing was one of the first to focus exclusively on machine intelligence. Turing begins that paper with the claim, "I propose to consider the question 'Can machines think?'" Instead of following the traditional approach to such a question by defining both of the terms "machine" and "intelligence," Turing replaces the question with a new one, "which is closely related to it and is expressed in relatively unambiguous words." He changes the question from "Do machines think?" to "Can machines do what we (as thinking entities) can do?" The advantage of the new question, Turing argues, is that it draws "a fairly sharp line between the physical and intellectual capacities of a man."

Turing also proposed a test inspired by a popular party game of that time, known as the "Imitation Game." In that game, a man and a woman would go into separate rooms and guests would then try to tell them apart by writing a series of questions and reading the printed answers sent back. In this game, the object for both the man and the woman is to convince the guests that they are the other person.

Turing described his version of the game as follows:

> "We now ask the question, 'what will happen when a machine takes the part of A in this game? Will the interrogator decide wrongly as often when the game is played like this as he does when the game is played between a man and a woman?' These questions replace our original, 'Can machines think?'"

Later in the paper, Turing also suggests an "equivalent" alternative structure that introduces a judge conversing only with a computer and a man.

In 1952, he proposed the version of the Turing Test that is more generally known today. In this final version, several individuals replace the single judge and ask questions of a computer. The role of the computer is to then make a significant proportion of the group believe that it is really a human.

The Turing Test is now considered as successfully passed if a computer is mistaken for a human more than 30% of the time during a series of 5-min keyboard conversations.

As a measure of how far along we are, a computer program called "Eugene Goostman" was said to have passed the Turing Test at an event with a total of 30 judges in June 2014 organized by the University of Reading.[4]

The university's School of Systems Engineering organized the event, and at that session, 33% of the judges at the Royal Society in London were convinced that it was a human on the keyboard at the other end of the link. The event was organized in partnership with RoboLaw, an EU-funded organization with interest in the regulation of emerging robotic technologies.

As would be expected in any such claim, there is still debate regarding the precise definition of the test itself, as well as the interpretation of the results. Nevertheless, the results do indicate a milestone against which other results can be compared.

However, the issue is much bigger than having a machine with a similar level of intelligence as a human. Nick Bostrom expresses this concern in his article "When Machines Outsmart Humans."

> "There is a temptation to stop the analysis at the point where human-level machine intelligence appears, since that by itself is quite a dramatic development. But doing so is to miss an essential point that makes artificial intelligence a truly revolutionary prospect, namely, that it can be expected to lead to the creation of machines with intellectual abilities that vastly surpass those of any human. We can predict with great confidence that this second step will follow, although the time-scale is again somewhat uncertain. If Moore's law continues to hold in this era, the speed of artificial intelligences will double at least every two years. Within fourteen years after human-level artificial intelligence is reached, there could be machines that think more than a hundred times more rapidly than humans do."[5]

In addition to conversational skills and information, yet another pertinent question is whether a machine can have emotions. This question begins to intrude into the area of religious philosophies, since one definition of "emotions" is based only in terms of their effect on behavior or on how they function inside an organism. In that case, emotions can be viewed as a mechanism that an intelligent being—or device—uses in order to maximize the utility and selection of its actions.

Given this definition of emotion, Dr. Hans Moravec, author of *Robot: Mere Machine to Transcendent Mind* in 1998, wrote, "Robots in general will be quite emotional about being nice people." Moravec argued that if fear is defined as a source of urgency, and empathy is acknowledged as a necessary component of good human-computer interaction, he concluded that robots "will try to please you in an apparently selfless manner because it will get a thrill out of this positive reinforcement. You can interpret this as a kind of love."

Daniel Crevier, Canadian researcher and entrepreneur focused on artificial intelligence and author of *AI: The Tumultuous History of the Search for Artificial Intelligence*, concluded that

---

[4]http://www.bbc.com/news/technology-27762088
[5]http://www.nickbostrom.com/2050/outsmart.html

"Moravec's point is that emotions are just devices for channeling behavior in a direction beneficial to the survival of one's species."

However, emotions can also be defined in terms of their subjective quality of what it feels like to have an emotion. The question of whether the machine actually feels an emotion, or whether it merely acts as if it is feeling an emotion is the philosophical question of "can a machine be conscious?" but in another form.

Nevertheless, artificial emotion—or AE—is an underlying aspect of the concept of AI.

The evolution of sensation in the animal world is the root of what are called "emotions" or "feelings." Without any construction capable of organic neural sensation, any synthetic/artificial/android/mechanical AI constructs will be unable to "feel" anything, or register this particular aspect of what "emotions" arise from in the same sense, as a human can understand those motivations.

However, the simulation of "emotions" can be managed superficially in a mechanical device. Synthetic/artificial/android/mechanical AI constructs can be programmed to falsify a smile or tears or any other physical expression of "emotion." Such a device could be designed to present all outward appearances of emotions. These devices will have no emotional "feeling," as such, regarding these "behaviors," since there is no nervous system at work, and no evolution of instinctual "affects" to register as a basis upon which their programmers can mimic human "emotions."

In that sense, artificial emotions are often considered to be the more dangerous aspect of the field of artificial intelligence rather than the more frequently discussed concerns with the potential physical threat inherent in growing machine "intelligence." AE is considered by many as a simpler method of gaining human empathy and establishing a foothold in human societies, since the human species is already comfortable with the concept of this "imaginary friend" through millennia of children playing with dolls and toys while attributing "emotions" to them in fantasy relationships. Similarly, AIs, such as the previously introduced Pepper, which were specifically made to seem "personal" or "caring," can bypass the natural skepticism of humans much more easily than AIs trying to simulate "intelligence." As the care-giving AI constructs gain more and more "emotional" value to people as "companions," "caregivers," "sexbots," etc., it may become harder to maintain control of their spread and acceptance into the human realm.

The question of whether highly intelligent and completely autonomous machines would be dangerous has also been examined in detail by futurists. That question is also an obvious element of drama and has also made a popular theme in science fiction, and has presented many different possible scenarios in which intelligent machines pose a threat to mankind.

However, one recurring topic is the possibility that machines may acquire the autonomy and intelligence required to be dangerous at a rate too rapid for human intervention or prevention. Some scientists therefore advocate the design of friendly artificial intelligence, meaning that the advances, which are already occurring with AI, should also include an effort to make AI intrinsically friendly and humane. Additionally, several tests have therefore been proposed for which to measure the progress of the development of human attributes in machines.

There is a very tight connection between Dr. Makimoto's third wave of technology and this scenario. Dr. Makimoto forecasted that the market value of robotics would exceed that of PCs and wireless communications devices as the primary application to drive the development of next-generation semiconductors. As we have seen in the growth of personal computers over the past decade, being in one of the leading positions in that growth cycle brought tremendous value to some companies.

Catching the next technology wave of robotics will likely have a similar impact, and could stimulate a technological surge in investments that would easily re-arrange the GNP rankings of those countries making the best technical investments. In addition to the financial possibilities, there is still the challenge of integrating those technologies into the social structures and political objectives of each country.

The question of whether robotics will be an extension of the growth of existing technologies or the beginning of a new slope in measuring the potential growth in computing power is a separate question. The relevance of that question is contained in the estimate that the development of AIs and other features are still at a starting point.

Dr. Anthony Berglas presented a paper in January 2012 that reviewed the technical progress in the development in artificial intelligence. He had concluded that newer computers are already almost powerful enough to host an artificial intelligence, and proposed that the production of more powerful computers should be limited in the future.

> "One frightening aspect of an intelligent computer is that it could program itself... At first only small improvements might be made, as the machine is only just capable of making improvements at all. But as it became smarter it would become better and better at becoming smarter. So it could move from being barely intelligent to hyper intelligent in a very short period of time. (Vinge called this the Singularity.)"[6]

Rather than attempting to make too fine of a distinction in the exact date, a broader conclusion might be that it is likely within the lifetime of the children of today they will encounter a robot that has roughly the same intellect and perhaps the same emotional capacity as a human teenager of today—which may be an unsettling thought for any parents of teenagers.

---

[6]http://berglas.org/Articles/AIKillGrandchildren/AIKillGrandchildren.html

CHAPTER 13

Brain-Machine Interface (BMI)

The previous chapter focused on social elements of how humans might integrate external devices such as highly skilled robots and computers. As the potential for linkage between ever smaller and more complex devices continues to grow, the integration of advanced technologies and humans likewise continues to accelerate.

Most of us today are familiar with the fact that many of the critical events associated with taking off and landing huge aircraft are primarily under the control of computers. The complexity and number of critical instrument readings has surpassed the ability of human pilots to predictably and accurately respond to that much information in a timely manner. The role of the cockpit crew in larger aircraft becomes more akin to managing many of these events at an executive level.

Similar activities are taking place in other professions. Villagers in some remote locations in South America that need medical attention often rely on computers for their initial interview and medical triage. Because of the remoteness of the location, their initial medical screening is done by a voice-activated computer system that can boast of a success rate higher than that of human medical screeners. The computer's diagnosis is the equivalent of the combined knowledge base of the entire group of doctors that contributed to the structure of the diagnostic decision tree. Nothing is overlooked, and the latest information is quickly integrated.

Progress also continues to be made by researchers in the medical field who are designing brain-controlled prosthetic limbs for people. Researchers have established that the brain controls the movement of our limbs by sending electrical signals to our muscles through nerve cells. When limb-connected nerve cells are damaged or a limb is amputated the brain still produces and trans-mits those motion-inducing signals even if the limb no longer exists. In recent years, researchers have created brain-machine interfaces (BMIs) that can pick up those interrupted electrical signals and redirect those signals to control the movements of a computer cursor as well as real or virtual prosthetics.

That integration between computers and humans is also accelerating in other areas. The ability to encode some elementary brain signals and transmit those signals to a remote biological subject (i.e., either animal or human) who then carries out those simple commands has also already been demonstrated.

The goal is to expand the amount of information that can be transferred along that Mankind to Machine Interface (MMI) or Brain-Machine Interface (BMI).

Much of this groundbreaking work has been done by Miguel Nicolelis at Duke University in Durham, North Carolina. In 2011, Nicolelis announced that he intended to develop a robotic, thought-controlled "exoskeleton" that will allow a paralyzed person to walk again. His target for demonstrating the current level of progress was the development of a robotic body suit that enabled a paralyzed person to kick a soccer ball during the opening ceremony of the 2014 Brazil World Cup. In actuality, while the verb "kick" may be too strong for what was demonstrated at this particular event, the state of the art was demonstrated by the ability of that patient to push the ball forward.

These events are already being reported against a backdrop of computing advances that are doubling every two years. Examples abound that further suggest knowledge and technology are still expanding at such an accelerated rate to eventually exceed the human ability to absorb that knowledge in a timely manner based on traditional learning techniques.

One of the most prominent examples is the work being done with the direct communications channel between a computer and a human brain. In 1996, Jan Scheurermann was a healthy 36-year-old mother of two children and running a small business in California when she suddenly developed an unusual illness. Her arms and legs slowly weakened to the point that she was confined to a wheelchair. She was eventually diagnosed with spinocerebellar degeneration, a condition that progressively deteriorates between the brain and the muscles. She relocated to Pittsburgh two years later where she has been working with the University of Pittsburgh School of Medicine and the University of Pittsburgh Medical Center to help develop technologies that could assist others dealing with quadriplegia.

In 2012, she was fitted with electrodes implanted in her brain. A computer that interpreted her mental instructions into the machine instructions that controlled a robotic arm read the signals from those electrodes. Within days of the final assembly, the mind-controlled robotic arm could help her feed herself, and she was eventually able to gain enough control to direct the equipment to serve a cup of coffee to her visitors.

In 2014, DARPA's Revolutionizing Prosthetics research group approached her with a different challenge. The Defense Advanced Research Projects Agency traditionally focuses on developing the latest in military technology. However, that role has shifted to other areas of defense, including advancements in traumatic brain injury (TBI) research and treatment.

DARPA director Arati Prabhakar said in a recent interview, "… in the next year, we have new areas growing and taking off in terrific directions: brain function research. We're now part of the presidential BRAIN Initiative … so 2014 is going to be an important year for those [research efforts] to get off the ground and for building the teams that are going to drive those areas forward."

"The focus on brain research is becoming increasingly critical as many troops return from assignments with TBI as well as psychological trauma from combat. DARPA is now involved in the development of techniques and devices that help restore active memory, while other efforts focus on deep brain stimulation and neuro-psychological research and potential treatments," she said in a recent interview.[1]

---

[1]http://fcw.com/Articles/2014/01/17/Prabhakar-profile.aspx

DARPA was interested in creating more responsive robotic limbs for injured veterans, and were searching for better methods to control these devices. Their hope was that the brain-interface technology could be pushed to a more complex level. The challenge was: could she use the same neural connections in a flight simulator to pilot an F-35 next-generation attack jet?

Prabhakar cited the breakthrough at the first annual Future of War conference in February 2015. "Instead of thinking about controlling a joystick, which is what our ace pilots do when they're driving this thing, Jan's thinking about controlling the airplane directly," She said. "For someone who's never flown—she's not a pilot in real life—she's flying that simulator directly from her neural signaling."

Prabhakar said the research is far from becoming reality. She also acknowledged that it raises fundamental moral and ethical questions about the intersection of biology and robotics. "In doing this work, we've also opened this door," she said. "We can now see a future where we can free the brain from the limitations of the human body and I think we can all imagine amazing good things and amazing potential bad things that are on the other side of that door."

And the reaction of the new pilot? She is definitely not combat qualified to fly the F-35, however she found her virtual piloting of a small Cessna plane around Paris and viewing the Eiffel Tower during the earlier training phase to have been "liberating."

CHAPTER 14

Acceleration Rate of Artificial Intelligence

At the same time that computers and robots gain in performance and skill sets, some evolutionary progress in human development can also likely be assumed. For example, dramatic changes in the rate of economic growth have occurred in the past because of some technological advancement. Based on population growth, the worldwide economy is thought to have doubled every 250,000 years from the Paleolithic era until the Neolithic Revolution. That new economy based on agriculture began to double every 900 years, which is a remarkable increase over the previous period. In our current era, beginning with the Industrial Revolution, the world's economic output doubles every fifteen years—sixty times faster than during the agricultural era.[1]

Robin Hanson (associate professor of economics at George Mason University and a research associate at the Future of Humanity Institute of Oxford University) argues that based on the previous periods, one might therefore expect future economies to double at least quarterly and possibly on a monthly basis if the rise of superhuman intelligence causes a similar revolution.[2]

Even if we accept the less aggressive approach of assuming that technological advances are likely to result in at least some level of disruption in the course of human intellectual development, how might this disruption come about as machines gain more intelligence and human attributes—and how close to that technological point of disruption have we already advanced?

Based on the information currently available, many technologists believe we are likely only 20–30 years away from robotics with the physical dexterity and level of information processing capabilities of a teen-aged humans—a thought that particularly some parents of teenagers may not find comforting.

As previously mentioned, Vernor Vinge has predicted four ways that a rapid expansion of intelligence could occur; super intelligent computers could "awake," large computer networks could develop a unique identity, the distinction between mankind and machines may become so tightly bound that the human may be considered to have acquired superhumanly intelligence, and biological sciences may improve natural human intellect.

In 2008, Dr. Anthony Berglas concluded that the speech recognition capabilities of a high-end computer were similar to that of a human, and that this activity seemed to require about 0.01%

---

[1] http://www.businessinsider.com/preparing-for-the-super-convergence-rise-of-the-bio-info-nano-singularity-2011-4

[2] http://anglejournal.com/article/2015-06-brain-emulation-economies-the-next-stage-of-human-development/

of the volume of a human brain.[3] If we accept that observation as an arbitrary starting point for measuring robotic computational capability relative to human capability, where does that lead us based on an estimated growth rate of technical advances similar to that of Moore's Law?

Table 14.1: Potential growth of computer intellect relative to human brain

Year	%	Year	%
2008	0.01	2028	10.2
2010	0.02	2030	20.5
2012	0.04	2032	41.0
2014	0.08	2034	81.9
2016	0.16	2036	163.8
2018	0.32	2038	327.7
2020	0.64	2040	655.4
2022	1.28	2042	1,310.7
2024	2.56	2044	2,621.4
2026	5.12	2046	5,242.9

Momentarily disregarding many of the parallel advances in other technological fields that supported Gordon Moore's observations, we can at least get an idea of how fast hardware progress could possibly take place under those assumptions of doubling of computational capabilities every 18–24 months.

The implications of these assumptions related to the sustainability of developing new technologies thus become clearer.

Additionally; if we compare our emotional responses to the left-hand columns and our reaction to the right-hand columns, we also discover another common human reaction. An almost natural reaction to the table above is that the early progression of the left-hand seems plausible, while the later progression on the right-hand side seem preposterous.

While the mathematic process of generating a geometric progression for all of the percentages is the same, our brain tends to be more comfortable working with smaller incremental changes. The power of a geometric progression of doubling over a fixed period of time only really becomes apparent as the numbers become of larger value.

However, since we are talking about the relative computational ability of a computer, we must also remember that computer efficiency can be improved in two ways. One method is to increase the speed at which transactions are completed, which is directly related to advances in hardware and the implications of Moore's Law regarding the processing efficiencies.

---

[3]http://berglas.org/Articles/AIKillGrandchildren/AIKillGrandchildren.html

The other method is to build the computer so that it can successfully perform more complex software tasks, as was highlighted by IBM in that company's shift toward cognitive computing systems. That trend is accelerating as the architectures of new high-end computers continue the trend away from the traditional von Neumann architecture[4] and toward cognitive structures which are much more similar to human thought processes.[5]

For example, we are already discovering other technology tracks, such as artificial intelligence (AI), that have already demonstrated the potential to make themselves more intelligent by modifying their own internal source code. These improvements would make further improvements possible, which would in turn accelerate further improvements, and so on.

This is a step beyond the allocation of resources and assignment of tasks, but not beyond the realm of consideration. This mechanism for an intelligence explosion within computers differs from an increase in computational speed in two ways. First, it does not require external effect: machines designing faster hardware still require humans to create the improved hardware, or to build appropriately efficient manufacturing facilities. An AI that is rewriting its own source code, however, could do so while still contained in an AI box or computing hardware structure.

There are naturally still some risks associated with this type of self-contained advance in intelligence. While not actively malicious, there is no reason to think that AIs would actively promote human goals unless they could be programmed as such. AIs might also compete for the scarce resources mankind uses to survive, such as a power grid.

Additionally, if the advance is so subtle or achieved so rapidly as to not be recognized, an AI might re-direct some of the resources currently used to support mankind to promote its own goals—potentially in conflict with human goals. The optimum structure of the AI may not be predictable or controllable under the self-improved design, therefore potentially allowing the AI to optimize for something other than was intended.

This second category of performance advance differs from computational power in that it is much harder to predict the outcome of the improvements. While speed increases seem to be only a quantitative difference from human intelligence, actual improvements in intelligence might be qualitatively different.

Eliezer Yudkowsky[6] observed the changes that advances in human intelligence have brought. He concluded that humans changed the world thousands of times more rapidly than evolution had done—and in totally different ways. He also concluded that the evolution of life had been a massive departure and acceleration from the previous geological rates of change, and improved intelligence could cause change to be as different once again.

Likewise, the possibility of unanticipated consequences leading to the development of technologies that have substantial social consequences also becomes a higher possibility—as recently demonstrated when almost all countries with strong technological bases in computing and communications services reluctantly acknowledged the extent to which communications between pri-

[4] https://en.wikipedia.org/wiki/Von_Neumann_architecture
[5] https://en.wikipedia.org/wiki/Cognitive_architecture
[6] http://rationalwiki.org/wiki/Eliezer_Yudkowsky

vate individuals were being monitored and recorded. As this is being written, various geographic sections are now pushing for stronger geographic isolation of Internet and data storage facilities in light of the public discussions of how the amount of cyber-snooping is being carried out. The possibility of cyber attacks against complete networks by an unidentified source or the hacking of private accounts is already a growing concern.

However, at the same time, advances in artificial intelligence and in creating more intelligent robotics are two of the most easily recognized technologies that are fundamental for accelerating the transition toward a Singularity-like explosion of technology. For example, some machines have in fact already acquired various forms of semi-autonomy, including the ability to locate their own power sources as well as to select targets to attack with weapons. In addition, some computer viruses that can evade elimination and have achieved at least an equivalent level of "cockroach intelligence."[7]

---

[7]http://www.nytimes.com/2009/07/26/science/26robot.html?_r=0

CHAPTER 15

The Industrial Revolution Revisited

Since many futuristic concepts are still under development, in some cases it is still too early to predict their practical—or selective—applications or even their possible date of commercial availability. Nevertheless, what has already been accomplished seems close to enough to the advanced possibilities that it is prudent to consider what impact such technologies might have on the worldwide social fabric. Regardless of how close the commercial applications of these possibilities are, it is not too early to anticipate the resulting disruption to existing social structures.

There is no doubt that recent technical advances have already had an impact on our daily lives. We appear to be constantly alerted to international security breaches, stolen passwords, and hacked computer accounts.

If we seek previous historic events on a larger scale that might give some forewarning of the level of social impact that can result from the rapid introduction of new technologies, examples abound throughout history—and none are more poignant or instructive than the period of the 1800s in Europe.

The tremendous changes in human conditions that occurred during the relatively short period of time of the Industrial Revolution were driven by technology advances in agriculture, transportation, textile, and metal manufacture. The benefits of that period have been significant and have brought remarkable improvements to mankind.

However, while this period thoroughly overwhelmed the old traditions in some cultures, the term "revolution" is somewhat misleading in the sense that it suggests an abrupt change. The substantial social changes, in fact, occurred gradually over the period of 1760–1860.

Nevertheless, as we approach the next phase of technical and industrial changes to be brought about by the range of new technologies on the horizon, the social challenges associated with those previous technological advances should also be acknowledged.

For example, while we are constantly being reminded of the jobs that are likely to be transferred to robotics and computers, the earlier period of the Industrial Revolution was also marked in Europe by the restructuring of rural labor during the decline of the feudal system in which the farmers were bound to the land that they cultivated. During that transition, scientific approaches to farming increased the yield of the fields and provided a greater surplus of food to support a larger non-agrarian population. Rather than keeping one of four fields fallow each year to replenish the soil, a four-crop rotation of the fields each year—based on wheat, turnips, barley, clover or alfalfa, and hay crops—also made it possible to support more livestock.

Other advances in agricultural techniques and practices also contributed to an increased supply of food and raw materials, while continuing changes in industrial organization and the arrival of other new technologies resulted in additional increases in production, efficiency, and profits.

Unfortunately, many smaller farmers often didn't have enough to survive as an independent farm even though everyone had the same share of land as before. Eventually the old system of farming was abolished and replaced with independent family farms. Beginning in the 16th century and extending to about 1820, many small farms were eventually abandoned as previous owners were forced to search elsewhere for work. In many cases, this restructuring of the labor pool directly supported the creation of new forms of labor as large factories were constructed.

Advances in the development of iron also contributed to the industrial transition. In 1720 most iron in England was imported due to the shortage of charcoal required for the smelting process, and iron was too expensive to use for large structures. This limitation was overcome in 1709 when Abraham Darby invented a method of smelting iron using processed coal (coke) instead of charcoal.

The iron industry took off after 1760 since iron ore and coal were both very plentiful in Shropshirel, England, and by 1777, work had begun for the first arch bridge in the world to be made of cast iron.

A new blast furnace built nearby had lowered the manufacturing cost of iron and encouraged local engineers and architects to address the long-standing local problem of crossing over the river.

Watt's steam engine appeared at around that same time (1774), along with numerous other advances, along within two decades newer high-pressure engines were being developed for transportation applications. Between 1750–1815, almost a thousand miles of improved roads were also constructed, which further reduced transportation costs by an additional 20–30%.

Following a contest in 1829 by the Liverpool and Manchester Railroad to test locomotives, railroads also began to spread rapidly.

One of the inventions with profound impact, the spinning jenny invented by James Hargreaves, was introduced between 1764–1770. The first big industry was cotton textile factories, though other kinds of factories developed as well. Prior to that, workers did piece work at home with spinning wheels and hand looms. What brought the workers together into a factory was the invention of machines for spinning that could spin more than one thread at a time combined with the use of water power for both the process of spinning and weaving the material.

The synergistic circumstance that made workers available was the previously mentioned advances in agricultural practices that had reduced the number of workers while increasing the amount of excess yields.

During the late 18th and early 19th centuries Great Britain became the first country to industrialize. With the synergy among these technical advances, the textile industry took off, and by 1833 the number of people employed in cotton textile factories in England had risen to 237,000.

Centuries before industrialization, children of poor and working-class families had always worked by helping around the house or assisting in the family's enterprise when they were able. Children frequently performed a variety of tasks critical to the family economy. However, relying on children to perform repetitive tasks for long periods of time in a factory environment became a new development. The labor force of the early textile industries was estimated at one point to consist of 46% women, and up to 15% children under the age of 13.

Great Britain eventually became the first country where the nature of children's work changed so dramatically that child labor became recognized as a social problem and a political issue. Significant new reform laws were first established in the UK through the Factory Act of 1833. The Factory Act forbid employment of children under the age of 9, and further restricted the hours of labor for children to nine hours a day for children of the ages 9–13 while allowing for labor of up to twelve hours a day for children aged 13–18.

Much of this transition to factories was driven by the public preference of cotton over wool. In the previous century, the government had already placed a ban on imported cotton goods because of the competition with the local wool and the linen industries. As a result, cotton goods were primarily imported from India during the second half of the 17th century.

However, cotton had become so popular that a "home-based" cotton industry had eventually been established. Bales of raw cotton were imported from the southern plantations in the American colonies. Due to the Atlantic trade winds, much of the cotton was first shipped to New England ports, and the corresponding ports on the west coast of Britain, such as Liverpool, Bristol, and Glasgow.

The local wool and linen manufacturers in the UK lobbied to place as many restrictions as possible on the import of cotton, but there was little they could do to stop the trend as cotton had become more fashionable, and by 1800, new cotton mills were constructed using the latest technologies. Spinning mules provided the very fine but strong thread that was used by the weavers on their power looms. The power looms themselves were operated by the latest designs in steam engines, using coal as the fuel.

Since the equipment and the raw materials needed in this new industry were far too expensive for the cottage industry structure, the cotton industry developed from a home-based cottage industry to a factory-based industry housed in cotton mills in less than one hundred years. The spinners and weavers no longer worked at their homes or at their own pace. They were now the employees of the person who owned the factory and who could pay for the raw materials. This new working class was now paid a wage to carry out a repetitive task in a cotton mill for a specific period of time each day.

The resulting social impact of this new form of labor required that many people needed to move into the areas close to where the cotton mills had been built. Due to the limited transportation and living options, in some areas 50% to 75% of worker families lived in a single room with no plumbing, and dumped their chamber pot into the street or gutter.[1]

The distress and discontent caused by these enormous changes also led to the resistance movement and protests of the Luddites. Taking the name of what was perhaps a mythical individual, Ned Ludd, who was reputed to live in Sherwood Forest, this loose organization of disgruntled workers and displaced farmers attacked factories and smashed the industrial machines developed for use in the textile industries. In a futile effort to save their traditional livelihoods, the men did not seem to have a national organization or any political agenda behind the Luddite riots as they attacked what they saw as the reason for the decline in their livelihoods.

In a factual event described by Charlotte Brönte in her novel *Shirley*, Luddites attacked William Cartwright's mill at Rawfolds near Huddersfield in April 1812. Cartwright and a few soldiers held the mill against about 150 attackers, two of whom were killed. An additional attempt was made on Cartwright's life the following week, and another manufacturer was killed by the end of the week.

In January 1813 at York three more protesters were charged with murder during a previous attack, and were found guilty and hanged. Fourteen others involved in the attack on Cartwright's mill or related activities were also hanged a week later.

In 1816, violence and attacks on machines broke out once again after a bad harvest and a general downturn in business. Luddites attacked another mill, smashing more equipment. Troops were called in to end the riots. Six men were executed and three others were transported out of the area.

In the meantime, with these technological advances in both spinning and weaving, it might be supposed that the supply of cotton would eventually become a limiting factor. However, technology was advancing in this area also. Eli Whitney's cotton gin made the extraction of the cotton fibers from the plant much easier, and the southern cotton growers were able to keep up with the demand for raw material.

And, of course, that source of raw material was based on slave labor.

Early in the slave trade, England had initially been a very active participant. However, in 1807 the same wave of social consciousness that limited child labor was also active in eventually banning the slave trade itself.

That was also the same year the Act Prohibiting Importation of Slaves was enacted in the United States, stating that no new slaves could be imported into the United States. At the next U.S. census in 1810, 1.2 million slaves were counted. Nevertheless, in 1860 on the eve of the Civil War, that number had swelled to almost 4 million.

---

[1] http://www.historylearningsite.co.uk/britain-1700-to-1900/industrial-revolution/life-in-industrial-towns/

Due to the continuing demand for cotton, some business elements within England still gave their support to the seceded U.S. southern states, as seen by their assistance in outfitting the Southern raider the C.S.S. Alabama. In its two years, this ship captured nearly a hundred federal merchant vessels.

After international arbitration endorsed the United States position in 1872, Britain settled the matter by agreeing to pay the United States $15.5 million for damages done by the Alabama and other warships built in Britain and sold to the Confederacy. (Senator Charles Sumner, the chairman of the U.S. Senate Foreign Relations Committee and a staunch supporter of the abolition of slavery originally wanted to ask for $2 billion in damages, or alternatively, the ceding of Canada to the United States.)[2]

The point of this historic diversion back to the social aspect of the Industrial Revolution is to suggest that as technology development continues to accelerate, there is likely to be a period of social experimentation before the new social impact is established. In particular, there is also the question of how those technologies will impact the worldwide social, economic, and worldwide organizational structures.

From a historic perspective, the Industrial Revolution brought significant growth in value along with rapid changes in social structures that few today would dispute or argue should be reversed. However, the transitional period was so rife with social displacement and change that any observer in the early 1800s would have had no way of anticipating the extent of social and technical changes waiting ahead.

In considering the social impact of the Industrial Revolution in light of recent forecasts of the future potential labor impact of some new technologies, we can also infer that our own current cultural structures will in fact once again experience a very vigorous period of experimentation.

For example, futurist Thomas Frey, author of *Communicating with the Future*, projects that over 2 billion jobs will disappear by 2030.

Posted by FutureSpeaker on March 21, 2014:[3]

"Indeed, as I've predicted before, by 2030 over 2 billion jobs will disappear. Again, this is not a doom and gloom prediction, rather a wakeup call for the world.

Will we run out of work for the world? Of course not. Nothing is more preposterous than to somehow proclaim the human race no longer has any work left to do. But having paid jobs to coincide with the work that needs to be done, and developing the skills necessary for future work is another matter.

Our goal needs to be focused on the catalytic innovations that create entirely new industries, and these new industries will serve as the engines of future job creation, unlike anything in all history."

---

[2] https://en.wikipedia.org/wiki/Alabama_Claims
[3] http://www.futuristspeaker.com/2014/03/162-future-jobs-preparing-for-jobs-that-dont-yet-exist/

Other sources have also suggested that the opportunities for employment will not be completely lost, and that there will always be requirements for human labor. However, without some level of prior planning, there is no guarantee that the physical locations of those new employment opportunities will be in the same location—or even in the same country—as the jobs that will be eliminated.

In the meantime, the debates continue over the metrics of how to measure the flow of new technologies, the rate at which advanced technologies can be integrated, new norms of personal privacy, and the future role of robotics. If there is any point of agreement, it is that there is no single broadly accepted metric by which we can predict the challenge ahead.

As mentioned earlier, the debate over quantifying technical progress and predicting how far into the future those trends extend will likely continue. What we have attempted to do in this book is to take two specific concepts, identify their current progress, and consider their social impact. The next two chapters present those two specific concepts.

CHAPTER 16

Singularitarianism

In the Introduction, we offered two concepts that would be useful in determining our current level of technological integration. One of those concepts was the merging of mankind and machine to the point that the distinction was meaningless.

We've also presented the European Industrial Revolution of the 1800s as a historically recent example of a technologically driven transitional period that triggered unanticipated and irreversible changes in human societies.

Both of those scenarios are often associated with the concept of Singularitarianism. Wikipedia defines Singularitarianism as follows:[1]

"Singularitarianism is a movement defined by the belief that a technological singularity—the creation of superintelligence—will likely happen in the medium future, and that deliberate action ought to be taken to ensure that the Singularity benefits humans.

Singularitarians are distinguished from other futurists who speculate on a technological Singularity by their belief that the Singularity is not only possible, but also desirable if guided prudently. Accordingly, they might sometimes dedicate their lives to acting in ways they believe will contribute to its rapid yet safe realization.

Time magazine describes the worldview of Singularitarians by saying that 'they think in terms of deep time, they believe in the power of technology to shape history, they have little interest in the conventional wisdom about anything, and they cannot believe you're walking around living your life and watching TV as if the artificial-intelligence revolution were not about to erupt and change absolutely everything.'"

Singularitarianism as a movement began in the early 2000s as a specific ideology when artificial intelligence researcher Eliezer Yudkowsky wrote his essay, "The Singularitarian Principles."[2] He stated that a "Singularitarian" believes that the Singularity is a secular, non-mystical event that is highly possible and eventually beneficial to the world, and that its adherents attempt to accelerate the safe and benevolent arrival of that technological Singularity.

"In June 2000 Yudkowsky, with the support of Internet entrepreneurs Brian Atkins and Sabine Atkins, founded the Singularity Institute for Artificial Intelligence to work

---

[1]http://en.wikipedia.org/wiki/Singularitarianism
[2]http://ieet.org/index.php/tpwiki/singularitarianism

towards the creation of self-improving Friendly AI. The Singularity Institute's writings argue for the idea that an AI with the ability to improve upon its own design (Seed AI) would rapidly lead to superintelligence. These Singularitarians believe that reaching the Singularity swiftly and safely is the best possible way to minimize net existential risk."[3]

Inventor and futurist Ray Kurzweil, noted for developing early speech and text recognition software, is also the author of the 2005 book *The Singularity Is Near: When Humans Transcend Biology*. Kurzweil's achievements in improving the ability of computers to understand speech and text are widely acknowledged. Statements in late 2013 indicate that he is "developing natural language understanding so that Google's computers can read text to the point of understanding the context in order to be more efficient in searching data."[4]

He defines a Singularitarian as someone "who understands the Singularity and who has reflected on its implications for his or her own life."[5] He estimates that the Singularity will occur around 2045.

There is also a strong connection between Dr. Makimoto's third wave of technology and Singularity. We can now see the dynamic technological trends developing that would bring about Dr. Makimoto's forecast of robotics overtaking PCs and mobile devices as the primary application to drive the development of next-generation semiconductors.

As we have seen over the past decade, the leading positions in that growth cycle brought tremendous value to some companies. Catching the next technology wave associated with robotics and very high performance computing devices will likely have a similar impact, and could easily re-arrange the GNP rankings of those countries that made the best choices in supporting this next technology wave.

In addition to the financial possibilities, there is also the unanswered question of socially integrating those technologies. The issue isn't the measurable rate of progress, but challenge to the existing social norm.

One current example of this challenge can be found in the ongoing discussions about how much electronic privacy should be provided to the public. It has been publicly exposed that government programs in many of the most technically advanced countries have been routinely accessing the email and phone conversations of private citizens. This discussion has most recently led to the publicly expressed opinion of some government officials that future hardware designs should be required to have backdoors to the processing elements.

Others have argued that any access that is intentionally left in the security of architecture eventually provides unauthorized future access.

During one discussion over the right for privacy and the right to have certain personal information removed from the Internet, the European Commission's director of fundamental

---

[3]http://en.wikipedia.org/wiki/Singularitarianism
[4]http://wired.com/2013/04/kurzweil-google-ai/
[5]http://singularity.com/qanda.html

rights caught the attention of the press as he referred to having the right to restrict what personal information could be exposed on the Internet by exclaiming, "You don't have the right to see me naked!"[6]

Other recent demonstrations of the risks include the capability to hack into the control network of a vehicle and take control of critical elements while that vehicle is in motion.

Without passing judgement on the complete philosophy of Singularity, it is therefore useful to notice where we already are along the continuum toward a new social structure, and the selection of the Industrial Revolution is useful as a previous example of social evolution.

In the transition from feudal social structures with home- industries toward the mechanization and repetitive tasks of the Industrial Revolution in Europe, a revision of the definition of work is very apparent. For example, the seasons and the weather conditions no longer determined the convenience or necessity of which tasks need to be done. Factory workers were "on the clock" and tasks were predetermined.

While the rural economy relied on physical labor, factory production placed more reliance on the force multipliers of the machinery. The result was a disruption in the traditional social structure: women and children could perform factory tasks, resulting in higher male unemployment; the close proximity of living quarters for the large groups of laborers—many with no running water—resulted in degrading and unhealthy living conditions. That level of dislocation lead to riots and attacks on factory equipment as well as physical assaults by unemployed men against some of the factory owners. Armed clashes between out-of-work men and security forces resulted in the arrest and hanging or deportation of many of the demonstrators.

Let's also recall that the new definition of labor also extended to the primary sources of those raw materials—particularly the primary source of the cotton. It is estimated that by 1860, there were roughly 4 million slaves in the country that was the major supplier of that cotton—the former British colonies that had formed the United States of America.

Arbitrarily assuming that someone born in the U.S. in 1860 might have had a life span of 90 years, they would have experienced the passing of the 16th amendment that ended slavery, the migration to the land and resources of the western states, the impact of iron, steel, high explosives, and the beginning of the age of aircraft associated with WWI, followed by a second world-war that was highlighted by the rapid evolution of high-mobility of heavily armored tanks, high performance fighter aircraft, and punctuated by the explosion of atomic bombs on relatively undefended civilian populations.

On top of these concerns is another familiar topic—the definition of work and labor. New studies of the forecasted range of jobs that will be absorbed by robotics continue to show up on a regular basis. However, the impact of these job transfers to economies that are already suffering less than optimum employment, along with a continuing transfer of wealth to a smaller segment of the population, the probability of social turbulence and a period of social adjustment seems more likely.

---

[6]http://www.theregister.co.uk/2014/11/05/right_to_be_forgotten_eu_panel/

CHAPTER 17

The Noosphere

What has become clear in the recent feasibility demonstrations of brain-machine interfaces (BMI) is the fact that once the physical location of the brain signal is correctly identified, the necessary conversion and translation of simple commands can be translated into a kind of machine-level message that can also be transmitted through common communications networks. Such usable networks include corporate computer networks, the Internet, or public communications systems. The ability to maneuver an aircraft within a three-dimentional area—even if accomplished at this early stage in a simulator—demonstrates the current level of accomplishment in controlling mechanical devices through human thoughts. The remaining challenge is in the detection and proper translation of more complex brain-generated messages, and work continues on this aspect. While the degree of complexity associated with locating those communication linkage points within the brain is still quite challenging, nevertheless the transmission distance between the original signal from the human brain and the end device is irrelevant.

For example, in one set of ongoing studies at the University of Washington, researchers demonstrated a noninvasive human-to-human interface. One researcher was able to send a brain signal by way of the Internet that would stimulate the finger movement of the second researcher.[1] This program in August 2013 is considered to be the first demonstration of human-to-human brain interfacing. In that experiment, one subject sat in his lab wearing a cap that included electrodes that were hooked up to an electroencephalography machine that reads electrical activity in the brain. The other subject was across the campus fitted with a cap that contained a trans-cranial magnetic stimulation coil placed directly over the section of his cortex that controls hand movement. The team had set up a Skype connection outside of the vision of the two subjects so the two labs could coordinate activities.

After the initial setup, one subject viewed his own computer screen and played a simple video game. Whenever he was supposed to fire a cannon at a target in the video game, he imagined moving his right-hand while being careful not to actually stimulate any muscle movements. The other subject wore noise cancelling ear inserts and could not see the game on the computer screen.

Almost instantaneously when the first subject thought to "hit the 'fire' button," the remote subject involuntarily moved his right index finger to push the space bar on the keyboard, as if he were actually viewing the game. This subject compared the feeling of his hand moving involuntarily to a nervous tic.

---

[1] http://www.washington.edu/news/2013/08/27/researcher-controls-colleagues-motions-in-1st-human-brain-to-brain-interface/

In another example of BMI, a new skill was acquired that was never before known to a species, in this case rats. Rats share a limitation with humans that neither species can detect the presence of ultra violet (UV) light. However, a professor at Duke University attached a UV detector through the skull and into the brain of a rat. When UV light was directed at a spot in a completely dark environment, the rat was trained to connect the tingle in a particular region of his brain resulting from the circuitry detecting the intensity of the UV light with the proximity of food. While the rat couldn't physically "see" the UV light in the normal definition, the rat did acquire the ability to "sense" the presence of that invisible light.

Researchers were also able to stimulate brain-to-brain communications between two similarly modified rats. The surprising aspect of the experiment is that when the trained rat was electronically connected to a second similarly modified rat, the trained rat appeared to be able to pass some aspect of his knowledge over the electronic channel to the newly modified rat.

Furthermore, the original modified rat in that experiment appeared to alter his own thought waves in order to assist the learning process in the electronic transfer of that new skill set to a second rat.[2]

In terms of human-to-human transmissions, there is already a philosophy which anticipates a similar level of communications among humans. Wikipedia defines the noosphere (sometimes noösphere) as a concept presented by Vladimir Vernadsky and Teilhard de Chardin to denote the "sphere of human thought…For Teilhard, the noosphere emerges through and is constituted by the interaction of human minds."[3]

One of the fundamental aspects of the noosphere concept deals with evolution. In the early 1900s Henri Bergson's L'évolution Créatrice[4] was one of the first to propose that our human evolutionary process is "creative" and cannot necessarily be explained solely by Darwin's theory of evolution by natural selection. L'évolution créatrice theory is supported, according to Bergson, by a constant vital force which animates life and fundamentally connects mind and body. In 1923, C. Lloyd Morgan took this work further, elaborating on an "emergent evolution" which could explain increasing complexity, including evolution of the mind. Morgan found that many of the most rapid changes in complexity of cell structures in living things have not been continuous with past evolution, and therefore did not necessarily take place through a gradual process of natural selection. He concluded that evolutionary experiences are also capable of jumps in complexity.

Others have also concluded that the complexity of human cultures, particularly languages, facilitated a quickening of evolution in which cultural evolution occurs more rapidly than biological evolution. In this aspect, the spread of information and ideas sounds very similar to the "contact theory" of the spread of ideas discussed in an earlier chapter.

Within these concepts, the noosphere concept proposes that knowledge and evolution have grown in step with the organization of the human population in relation to itself as it populates

---

[2] http://www.nytimes.com/2013/03/01/science/new-research-suggests-two-rat-brains-can-be-linked.html?_r=0
[3] http://en.wikipedia.org/wiki/Noosphere
[4] https://en.wikipedia.org/wiki/Creative_Evolution_(book)

the Earth. As mankind organizes itself in more complex social networks, the higher the noosphere will grow in awareness. Teilhard argued that the noosphere is growing toward an even greater integration and unification, culminating in the Omega Point—an apex of thought and consciousness that he saw as the ultimate goal of humanity.

In describing this Omega Point as a future time beyond which it is impossible to anticipate the impact on our social structures and human experiences, the human impact is somewhat analogous to the Singularity—but with a different driving force.

In a previous chapter, we identified two dominant scenarios associated with Singularity events beyond which it is difficult to anticipate future events. The first scenario represented by the ability of a robot to perform functions normally restricted to humans has been discussed in the previous chapters.

The other scenario is the advancement of human intellect and capabilities by way of technologically assisted enhancements. Simplistic examples of this second noosphere scenario of events that might trigger a more rapid evolution of the human brain might range from a construction worker inside a remote cab of a tractor and surrounded by screens controlling a back hoe at a distant site, or perhaps a more complex target of a paraplegic kicking a soccer ball, as well as the use of brain waves to control a mechanical device such as a jet aircraft or a drone.

Additional potential examples have also been suggested. One of the most commonly identified possibilities are individual brain inserts that could either augment previously damaged brain segments or control prosthetic devices—or eventually to simply enhance the capabilities of otherwise healthy cells and organs.

This second scenario of Singularity combining humanity and machinery also suggests the ability to augment human mental capabilities over a wide range of possibilities. One area of potential advancement of human intellect is the ability to literally merge the brainpower of several humans. If, in fact, Bergson is correct that there is an evolutionary "creative" element that stimulates and accelerates human evolution at a faster rate than explained solely by Darwinian natural selection, the most direct communications channels between humans might accelerate that fundamental process of human evolution to a higher plane.

In the sense of accelerating the evolution of the human brain by exposure to more information, the rate at which we can understand and absorb new information is a definite limiting factor.

The nosphere concept is based on the theory that a method will eventually be developed which will allow a faster brain-to-brain transfer of information than our historic senses of sight and sound. The challenge is the ability to translate information into some format that can be transferred at a faster-than-normal rate to another physically remote subject.

If that challenging task can been completed, the previously discussed experiment with the rats comes into play. The primary rat appeared to be able to transfer—or at least facilitate the learning of—its knowledge of the relationship between UV light and food by way of electronic transmission to another rat. Knowledge was being converted into electrical signals that were transmitted over a physical communications channel to the correct reception point in another subject.

While this particular experiment included hard-wired brain inserts, other recent human experiments have also been able to achieve results by way of non-invasive external monitoring of alpha waves that can be sensed at skin levels, such as the ability of one subject to control the involuntary actions of another subject over an Internet connection.

As previously noted, what has become clear from the cross-campus transfer of thoughts related to finger impulses is the fact that the distance between the original signal and the end device is not a factor. Once the brain signal is recognized, it can be translated into the kind of machine-level message that can be transmitted through normal communications networks or over the Internet—or potentially stored for later transmission.

If we accept the possibility that the transmitting rat consciously changed its brainwave pattern in the process of transferring information to the receptor rat, then a group linkage of human minds appears to be possible. The question then becomes whether such an event would trigger such a rapid evolution of those connected human brains that the outcome would be unpredictable—the Singularity.

CHAPTER 18

Mapping the Brain

We have identified the progress in two separate scenarios associated with technology-assisted advances beyond which it is difficult to anticipate future events.

The first scenario is the advancement of human intellect and capabilities by way of technologically assisted enhancements—the intellectual merging of human and machine capabilities to the extent that the distinction between man and machine is no longer meaningful. The second scenario is the stimulated acceleration of human intellect resulting from direct brain-to-brain linkage. As previously mentioned, those two scenarios do not directly overlap.

The one area where they do merge is in the mechanical linkage to the human brain. Brain waves have already been utilized to control the mechanical augmentation of limbs for physical activities, and one aspect of singularity is the concept of "brain inserts" that can facilitate the transfer of knowledge or skill sets to the recipient. In a similar sense, noosphere concepts also assume that brain-to-brain connections will accelerate human evolution.

This point of view can extend to a vision of technologically enhanced intellect also contributing to a potential elevation to a higher spiritual level.

Dr. Mori, who established the "Uncanny Valley" which was discussed in a previous chapter, expressed in *Buddha in the Robot* an Eastern religious perspective of what may ultimately be a distinguishing characteristic of the development of advanced technologies as assistants and enhancements to humans.

Taking on additional intellectual capabilities by physically merging with technically driven enhancements can be seen as an evolutionary elevation to a higher level of consciousness and spirituality which is more in line with some Eastern religions—a point of view that is not found within the major Abrahamic religions.

Nevertheless, while still in an early stage of development, more sophisticated human-to-machine interfaces are already well established in terms of the technical feasibility. For example, Google Glass is reported to have the capability of using physical movements, such as a nod or a single wink, to issue logic commands.

IMEC, Holst Centre, and Panasonic have also unveiled a new version of their wireless EEG headset. The device monitors brain waves using eight channels and transmits readings directly to a wireless receiver up to 10 meters (30 feet) away. Because it can be applied by the user without assistance from a highly trained professional, coupled with the fact that it can be used outside the clinic, the device may open new doors for using EEG to study the brain in a variety of situations and environments.

Figure 18.1: Photo 1.

At a higher level of interface, researchers in Samsung's Emerging Technology Lab are already testing tablet computers that can be controlled by one's brain.[1] This particular program uses a cap that is studded with monitoring electrodes to transmit impulses from the surface of the head to the computer. These types of programs typically rely on alpha waves that can be detected close to the surface of the body.

A more invasive method is also under study with the goal of aiding paraplegics. In these programs, brain waves can be utilized to control the mechanical movement of limbs for physical activities.

We have already introduced DARPA's interest in brain-controlled prosthetics, and much of that motivation is for combat-related injuries. While much of that agency's focus will remain on traditional programs, emphasis is also shifting toward advancements in research and treatment of traumatic brain injuries. DARPA's brain research focuses on troops who return from combat with traumatic brain injuries as well as psychological trauma from combat. Some of DARPA's work also involves development of techniques and devices that help restore active memory, while other efforts focus on deep brain stimulation and neuro-psychological research.

This capability to control external devices results from the fact that the brain can still create and send control sequences even if the limb is not responding—or no longer there. The channel's logical interface challenge is to pick up the exact brain location generating those signals, detect, and transmit that signal.

One of the leading programs in this area of research is "BrainGate," pioneered by Dr. Donoghue. BrainGate participants have previously demonstrated neutrally based two-

---

[1] http://www.technologyreview.com/news/513861/samsung-demos-a-tablet-controlled-by-your-brain/

Figure 18.2: Photo 2.

dimensional point-and-click control of a cursor on a computer screen and rudimentary control of simple robotic devices.

The pilot for a follow-on clinical trial, dubbed BrainGate2, employs the BrainGate system initially developed at Brown University, in which a baby aspirin-sized device with a grid of 96 tiny electrodes is implanted in the motor cortex—a part of the brain that is involved in voluntary movement. The electrodes are close enough to individual neurons to record the neural activity associated with intended movement. An external computer translates the pattern of impulses across a population of neurons into commands to operate assistive devices.

Even that physical connection on a chip inside the head could soon vanish as we gain a much greater understanding of the brain, along with a much higher level of complexity in the technologies that empower brain-computer interfaces. This program would enable a much higher data rate exchange between human and computer.

While these examples demonstrate the ability to regain and enhance human physical capabilities, there is evidence that adding new mental capabilities may also ultimately be possible.

In order to expand these capabilities, the challenge is to locate the specific three-dimensional location within the brain of that particular signal itself. Once the particular signal can be detected, then it can be used to control an electro-mechanical device. In that respect, the recent Brain Activity Map project initiated by the current Obama administration is a decade-long research project that aims to build a comprehensive map of the brain and its control centers. This project is considered crucial to expanding the potential of brain interfaces.

CHAPTER 19

Conclusion

We have presented in this book the current levels of achievement in such diverse technologies as artificial intelligence, robotics, electronic physical integration between humans and machines, as well as brain-to-brain electronic connections.

We have also presented arguments and some details related to ongoing debates about the rate at which knowledge and technologies are increasing, whether the arrival of new information is at a faster or slower rate than in the past, and in particular whether that rate of advance of new technologies will eventually trigger a social adjustment on the order of the Industrial Revolution. Our objective has been to demonstrate current examples of technical progress in the area of data processing and integrated circuits. Continuing the technical progress will likely cause a social restructuring of a magnitude at least similar to Europe's Industrial Revolution of the 1800s.

Many leading technologists believe that such an event will likely be upon us within the lifetime of the majority of people alive today.

We have also suggested mile markers that can help us estimate the speed at which we are approaching what some technologists are calling a "period of turbulence."

Another set of markers that can be easily checked over time is the forecasted number of human jobs that will be replaced or restructured alongside the current and projected trends associated with the transfer of wealth relative to the population within the various industrial countries.

There may be no ideal mathematic relationship between these two charts, but there may be some correlation relative to the rate at which we are approaching a higher level of social discord.

We also have presented forecasts from various technologists that anticipate substantial advances in technologies within the next twenty years, and we have attempted to share the sense of exhilaration that accompanies the rapid expansion of technologies. Part of that excitement comes from the sense that history is being created in our lifetime and that the future course of civilization may be shifting.

However, the purpose of this book is to call attention to the additional social challenges that are likely to be ahead. The ultimate challenge to mankind is thus the ability to successfully integrate that much technology and information into our cultures and societies. Regardless of the rate at which it appears, the point is to draw attention to the relationship between new technical progress and the potential socio-economic impact of those technologies.

The common element among the various technologies is the acknowledgement that something fundamental to human existence has changed.

One recurring topic is the impact that robotics will have on human employment opportunities. Regardless of the net impact advanced technologies will have in this area, the one issue we cannot ignore is that we are once again—as previously occurred during the Industrial Revolution—changing the nature of work and the value of labor.

In Dr. Makimoto's previously mentioned book *Digital Nomad* published in 1997, an additional thought-provoking topic was also raised regarding the relationship between governments and their citizens.

In an earlier chapter, we quoted Dr. Makimoto's opinion that the world-wide channels for communications have created almost unlimited opportunities for individuals to exchange ideas and establish new relationships.

Today we can see several sides of this topic. Many of us today are in constant communications with individuals in other parts of the world through business or personal interests. At the same time, we frequently read of radical organizations that have demonstrated their ability to recruit followers from countries far away from the actual battlegrounds of the organizations.

On the other side of the argument is the topic of personal privacy as governments are insisting that they be given backdoor access to all channels of personal communications and computing equipment in order to maintain the safety of national organizations.

However, at the same time, we cannot ignore the conditions reported in a small province in India, which reportedly levied fines on any young women using cell phones. Nor can we overlook the accounts of the impact of the Internet and international communications channels credited to contributing to the Arab Spring uprisings, or the growing concern about the privacy and safety of the Internet itself.

By the same token, we also cannot ignore that our innate thirst for knowledge will inevitably drive us forward regardless of whether we believe in a linear or a geometric increase in knowledge.

In that sense, we can appreciate the tremendous advances in the field of communications and marvel at the potential of combining snapping algorithms with altered reality concepts and applications on our cell phones.

We have included some examples of previous technological events that have had a very strong and unanticipated social impact. We also presented our so-called "San Francisco Model" to demonstrate that the initial usage of any new technology often becomes modified over time to reflect the continuing evolution and dynamic tension between what *can* be done and what *should* be done to better meet the broader needs of the society. We have little control over the rate at which information is advancing. The only real debate is the rate that we can absorb those technologies into our social and political structures.

"Arnold Toynbee (14 April 1889–22 October 1975) was a British historian and research professor of International History at the London School of Economics and the University of London, author of numerous books, and specialist on international affairs. He was one of the most read scholars in the 1940s and 1950s."[1] He is best known for his 12-volume *A Study of*

---

[1] http://en.wikipedia.org/wiki/Arnold_J._Toynbee

History (1934–61), through which he studied the rise and fall of 26 civilizations in the course of human history. In his 1972 abridged edition of *A Study of History* perhaps he best expresses the seriousness of this technological challenge that is still ahead.

"The challenge presented by the introduction of a new force into the domestic life of a society cannot be ignored with impunity. In the new situation that the new event has produced, social health can be preserved only through the adjustment of the old pattern to accommodate the new element; and this adjustment is tantamount to a replacement of the old pattern by a new one, or, in other words, to a thorough-going reconstruction of this particular social universe. The penalty for ignoring the necessity of making this new adjustment, or for seeking to evade it, is either a revolution, in which the new-born dynamic force shatters a traditional culture-pattern that has proved too rigid to adapt to it, or else an enormity engendered by the introduction of the new force's driving-power into the structure of an obstinate culture-pattern whose fabric has proven tough enough to withstand the new force's unprecedented powerful pressure. The encounter between a new culture-element and an old culture-pattern is always governed by the same set of circumstances, whether the new element happens to emerge from within or to impinge from outside. In both these variations on a situation which is ultimately the same in both cases, the introduction of the new element condemns the old pattern, *ipso facto*, to undergo a change in either its structure or its working. Unless this inexorable summons of new life is effectively met by an evolutionary adjustment of the culture-pattern's structure, the visitant, which in another context is either harmless or even creative, will actually deal deadly destruction."

The future challenge of the Digital Revolution is certainly not in the technical achievements, or even the precise rate at which technology is expanding. The future challenge will be in the ability to manage the social disruption as we absorb and integrate so much new knowledge into our societies and daily lives—even as those technologies are transforming our social and political structures.

How much more stress will fall on our various cultures if we eventually have some humans with brain inserts to assist their memory and thought processes, bio-mechanical parts to replace damaged limbs and boost human performance beyond any previously known limits, brain-to-brain and brain-to-machine exchanges of information over the Internet, industrial robots that compete for our jobs, and domestic robots trained to respond to all our physical requirements and emotional moods?

The issue will no longer be the difference in lifestyles between those that have access to the newest technologies those who do not. The issue will be the chasm between those groups and cultures that aggressively assimilate advanced technologies into their core beliefs as opposed to other groups that either do not have that opportunity to participate or have rejected those new technologies because the implications are simply too disruptive to reconcile with their beliefs.

Will that technological chasm be too wide for some cultures to traverse? Will it spark another response similar to the Luddite's in England's Industrial Revolution, who attacked the factories as symbols of their lost cultural identity and traditional economic opportunities?

Welcome to the Digital Revolution.

Author's Biography

BOB MERRITT

Bob Merritt is a 30-year veteran and recognized authority for semiconductor and communications markets. He is founding partner of Convergent Semiconductors, LLC. Convergent provides insightful market intelligence for memory and related technologies/applications specializing in illuminating the impact resulting from market and technological events and forecasting the opportunities created by those event changes. The topic of robotics is a technological, market, and social changing event.

Starting in the semiconductor industry as a sales engineer for Intel, Mr. Merritt's successes include the achievement of sales, revenue, and operational goals in marketing and operations at major semiconductor memory companies.

His consulting experience is global, providing insight to the World Bank on investments required for semiconductor memory and other semiconductors for an Asia Pacific market. He was also a co-author of an overall Memory Report for Credit Lyonnais Securities financial analysts in Asia and Europe. Clients worldwide include major participants in the semiconductor and electronics industry.

He has served as an expert witness in the San Jose Superior Court as well as the San Francisco Superior Court.

He was also previously under contract to EDN to write a bi-weekly blog on advanced and new memory technologies under the title "Professor Memory."

Printed in the United States
by Baker & Taylor Publisher Services